FUNCTION ANALYSIS

Systematic Improvement of Quality and Performance

FUNCTION ANALYSIS

Systematic Improvement of Quality and Performance

Kaneo Akiyama

Andrew P. Dillon, Translator

Productivity Press, Inc.
Cambridge, Massachusetts   Norwalk, Connecticut

Originally published by Japan Standards Association as *Kinoubunseki*.

Productivity Press
P.O. Box 3007
Cambridge, Massachusetts 02140
United States of America
Telephone: (617) 497-5146
Telefax: (617) 868-3524

Library of Congress Catalog Card Number:
ISBN: 0-915299-81-X

Book and cover design by Hannus Design Associates
Printed and bound by Bookcrafters
Printed in the United States of America

Library of Congress Cataloging-in-Publication Data

Akiyama, Kaneo.
[Kinō bunseki. English]
Function analysis: systematic improvement of quality performance / Kaneo Akiyama; Andrew P. Dillon, translator.
p. cm.
Translation of: Kinō bunseki.
Includes index.
ISBN 0-915299-81-X
1. Value analysis (Cost control) 2. Production management — Quality control. 3. Engineering design. I. Title.
TS168.A3513 1991 91-8681
658.15'52—dc20 CIP

91 92 93 10 9 8 7 6 5 4 3 2 1

Contents

List of Illustrations

From the Publisher

Four or five years ago, a dollar was worth 240 Japanese yen. Today, a dollar is worth only 130 yen — a decline of almost 50 percent. Throughout this devaluation, Japanese companies still sold their materials in the United States. And while these same Japanese companies couldn't very well double the prices of their automobiles or VCRs and remain competitive, they could — and did — put pressure on their engineers to reduce the cost of parts. They accomplished this ultimately by streamlining their manufacturing operations, by reducing the variety of parts while keeping up with the growing product diversification, and by carefully studying the functions of each part from the design stage on. In short, the Japanese reduced their costs because they had to.

In the past, U.S. industries were busy creating specialists — design engineers, quality assurance experts, mechanical engineers, and so forth — and establishing areas of domain. On one hand, this was good because it encouraged people to become in-depth experts in their areas. On the other hand, the resulting workforce hierarchies grew unmanageable.

Japanese companies turned away from the single-skill concept and disseminated information factory-wide. A multi-skilled workforce resulted that was informed enough to study, find, and resolve many of the problems faced on a daily basis. This heralded the birth of cross-functional teams.

In today's world class manufacturing environment, these concepts are no longer news. And, in fact, value analysis and its foundation, function analysis, are not news to most Americans. They were developed at General Electric in the 1940s. However, what is important now is the application and evolution of these two concepts. The Japanese, forced by circumstances to be innovative, borrowed the ideas from the Americans, who were still basking in the sun of industrial prosperity and happy to share. The Japanese quickly adapted the ideas to their needs.

Kaneo Akiyama's book *Function Analysis: Systematic Improvement of Quality and Performance* cuts to the heart of the matter. A professor and consultant in value engineering, he studied with Japan's top VE experts and understands it thoroughly. His book explains the need for function analysis and how it applies to quality control, process improvement and design, and product improvement and design. Tracing the history of function analysis, he presents a detailed explanation of two categories of FA techniques: function definition and function systematization.

Akiyama comments on the various types of systems — such as product systems, production systems, management systems, and service systems — that can be analyzed. He then applies function analysis to business aspects of product improvement and design, process improvement and design, quality control, and so forth and analyzes several case studies. The great value of this book lies in learning in depth about function analysis as a tool to improve our products and to reduce the cost of their manufacture.

Many people assisted in producing this book. In particular, I wish to recognize Andrew Dillon, translator; Cheryl Rosen, project editor; Christine Carvajal, copyeditor; Kathlin Sweeney, production and design; the staff of Rudra Press, artwork and composition; Dick Hannus, book cover design; and

BookCrafters, printing and binding. Finally, I offer sincere thanks to Mr. Masayuki Nishiie, Director General of the Japanese Standards Association, for kindly permitting us to publish this book in English.

Norman Bodek
Publisher

Foreword

In his book *Function Analysis: Systematic Improvement of Quality and Performance,* Kaneo Akiyama demonstrates once again the capacity of the Japanese to methodically dissect a methodology or process, eliminate the irrelevant, and restructure the salient elements into a more understandable and useful product. The overall value of the book is in describing the power of function analysis as an analytical tool. In doing so, Akiyama leads the reader through a variety of ways to perform function analysis.

Function analysis exemplifies processes conceived of by Americans, used more effectively by the Japanese, and then returned to the United States with a better appreciation for the original product. When Lawrence D. Miles created the *value analysis* concept at General Electric in 1947, he introduced function analysis as the basic foundation on which value analysis stands.

Miles understood that to significantly improve product value (that is, to increase a product's function while reducing its cost) we must first understand the product's function and the contributing functions of its individual components. He rightfully observed that when a function is not understood clearly we tend to use excessive and overly complex verbiage to express our understanding of that function. He said that a function that is understood clearly can be expressed simply by using an active verb and a measurable noun. Selecting the proper verb and noun, however, proves more difficult than it seems. In his book *Techniques of Value Analysis*, he writes:

> While the naming of functions may appear simple, the exact opposite is the rule. In fact, naming them articulately is so difficult and requires such precision in thinking that real care must be taken to prevent abandonment of the task before it is accomplished.

Miles was an active supporter and member of the Society of American Value Engineers (SAVE). Throughout his professional life he challenged others to play and experiment with the concept of function analysis. The result of this encouragement was the development of new models, variations, and applications for performing function analysis. "Random function determination," "numerical evaluation by paired comparison," and the "function family tree" are a few of the many variations that have evolved.

At SAVE's international conference in 1966, Charles W. Bytheway introduced a new concept he called the "function analysis system technique" or FAST. Bytheway's contribution was his recognition that functions do not exist randomly. To keep a system operating, functions depend on other functions in the same way component parts depend on other parts. Bytheway diagrammed the system using intuitive logic. By asking "how" and "why" the function in question is being performed, a functionally expressed network can be drawn to show the dependent relationship of one function to another. This simplifies the value analysis of entire systems. By (1) stripping away the complexities of product assemblies and components and (2) addressing their contributions to the performance of the system's prime or basic function in function terms, multi-disciplined task teams can use a common language to communicate effectively with each other and to resolve complex issues.

During this growth period, value analysis and function analysis expanded their fields of application from product cost reduction to product design and development, systems, procedures, organization effectiveness, productivity improvement, and into virtually every product, process, construction, and service industry.

While Akiyama's research has uncovered many approaches to function analysis, for his purposes the author has settled on a variation of the function family tree that is popular in Japan. This version is found in the United States, where it is called the "hierarchical" function model. Most practitioners tend to group all such models under the FAST label, however, a practice that causes confusion and academic debate regarding which approach is better.

My opinion as a practitioner is that the issues, problems, and opportunities under study should determine the function analysis version that is applied. Furthermore, it is my opinion that because of its basis on intuitive logic, the function analysis system technique and its acronym "FAST" should be reserved for its creator, Charles W. Bytheway. Nevertheless, regardless of semantics, Kaneo Akiyama logically presents the development and effective use of function analysis. This is a book that I must have on my bookshelf.

J. Jerry Kaufman

Preface

Function analysis came into being in 1947 as a technique for analyzing desired functions for designing and improving products. Since then, its domain of application has grown to embrace not only the "hardware" of product design and improvement, but also "software," such as processes and operations, and even service operations and indirect management activities. In the face of these expanded applications, we now see function analysis methodology as a collection of techniques for analyzing the aspect of function universally present in all systems.

Consider the concept of paradigms. Different management technologies have their own individual paradigms. The industrial engineering (IE) paradigm is to analyze work in great detail and then eliminate everything that is wasteful, irrational, and inconsistent *(muri, mura* and *muda).* The quality control (QC) paradigm is, of course, one of control, management, or constraint. The value engineering (VE) paradigm looks at things in terms of their values. Innovation — revolutionizing things — requires changing paradigms, and that is where the function analysis paradigm comes in.

Let us look at one example. Conventional management technologies have made little headway in penetrating the realms of service and indirect management activities. The ongoing structural shift from secondary to tertiary industries, however, is making improvement and innovation in such areas increasingly important. Both service operations and indirect management activities exist in order to carry out certain functions. Function analysis does two things for such activities. It makes them intelligible (and objectifies them) as systems, and it promotes more effective improvement and innovation.

From the point of view of actual activities in companies, even quality control still focuses largely on the manufacturing stage of production. The methods of quality control, moreover, have yet to break out of the manufacturing field. But hardware aspects of product quality are nearly all determined at the development and design stages. Applying function analysis to product development and design reveals the kind of quality customers demand in terms of product system functions and makes quality assurance achievable in the product development and design stage.

In "software" areas such as service and indirect management activities, the notion of quality is sometimes vague, and reliable techniques of analysis have yet to be established. However, here too, the use of function analysis reveals the nature of quality in terms of system functions and makes it possible to apply quality control to such areas. Function analysis, in other words, allows us to analyze the functions of a given system and to promote understanding of that system. At the same time, it provides an objective model of the system and serves as the starting point for improvement and innovation.

This book was conceived to facilitate the use of general function analysis in a variety of business management situations.

Chapter 1 introduces the conceptual approach and explains the significance of function analysis.

Chapter 2 traces the history of function analysis as developed by Lawrence D. Miles.

Chapter 3 considers the need for function analysis in process improvement and design, in service and indirect management activities, and in quality control.

Chapter 4 offers detailed explanations of two categories of function analysis techniques: function definition and function systematization.

Chapter 5 provides commentary on various types of systems that can be analyzed through function analysis.

Chapter 6 explains the application of function analysis to several aspects of business: product improvement, product design, process improvement, process design, support activity improvement, support activity design, and quality improvement.

Chapter 7 presents case studies of applied function analysis and summarizes postanalysis improvement and design processes as well as the content of improvement and design plans.

In November 1967, Sangyō Nōritsu Junior College published a book entitled *Function Analysis,* by Masatosi Tamai, a former professor emeritus who has since passed away. I first became interested in function analysis as the basis of value engineering around 1975, when I read Professor Tamai's book at the outset of my VE studies.

The methodology of function analysis, like nearly all management technologies, originated in the United States. Its arrival in Japan around 1955 sparked considerable interest, for here was an extremely metaphysical and lucid technique for analyzing the core of human behavior and thinking. That interest remains unchanged today, but it has not yet extended to Professor Tamai's *Function Analysis.*

Since Professor Tamai's death, I have continued to receive guidance in VE and function analysis from his successor, Hiroshi Tuchiya, himself a professor at Sangyō Nōritsu Junior College and director of the Value Management Institute at my home institution, Sangyō Nōritsu University.

In Japan today, the field referred to as VE or function analysis owes a tremendous debt to Professor Tuchiya and many others, including Yoshio Nakagami and Kazunari

Morioka, professor and assistant professor respectively at Sangyō Nōritsu University, and Fumito Nakano, manager of the Japan Value Engineering Association. Without the guidance of these mentors and of my many colleagues at the Value Management Institute, I would never have been able to bring this book to a conclusion.

I have also benefited from the cooperation of numerous companies for which I have served as consultant, and of industry participants in my function analysis seminars. Once again, I would like to extend my thanks to them.

Last of all, I would like to express my gratitude to Messrs. Iizumi and Kurata of the Japan Standards Association's Publishing Division for cheerfully grasping my aims and for undertaking the publication of this book. For the translation and publication of this book in English, I thank Norman Bodek and the staff of Productivity Press.

Kaneo Akiyama
March 1989

FUNCTION ANALYSIS

Systematic Improvement of Quality and Performance

1

What Is Function Analysis?

In Japan, the word for *function* (*kinō*) is rarely heard in everyday speech. Japanese dictionaries define function, as in function analysis, as follows:

- Working, action.
- The action of something. Any of the specific roles possessed by each mutually interacting element constituting a whole.

Since *kinō* is a translation of the English word *function*, it is perhaps instructive to look at the meaning of the word in English, according to my English-Japanese dictionary: Action, purpose, occupation, duty, mathematical correspondence.

THE MEANING OF *FUNCTION* IN FUNCTION ANALYSIS

In function analysis, the term *function* has three meanings.

Functions Are the Actions, or Workings, of Things

There are a number of ways to construe the term *things*. Material objects can be understood as static, fixed. Intangible things can be taken as perceptible phenomena. The term *function* in *function analysis* refers not to static, fixed, or apparent facts, but to the *actions,* or workings, of dynamic, interdependent, process-oriented things. It is these actions or workings that are analyzed in function analysis.

Functions Are Purposive

Actions can be ascribed to a variety of things. For example, there are various functions in natural phenomena, but the pursuit and study of those functions belong to the natural sciences. Function analysis concerns the functions of products or services. Products and services possess functions — actions useful to people — and companies intend that those actions serve people's needs or purposes. Function analysis analyzes functions in this *purposive* sense.

Functions Are Concepts

Things that are static, fixed, and apparent can be defined in terms of shapes and colors. This is not the case with functions, which are dynamic, relative, and process-oriented. Thus, functions are *concepts,* and the analysis of these concepts is function analysis.

THE MEANING OF *ANALYSIS* IN FUNCTION ANALYSIS

The field of management technology encompasses a variety of analytical techniques and practices, but the word *analysis* can be generally defined as:

- the breaking down of something to reveal the various components, elements, and aspects of which it is composed
- the division and identification of each of the various marks or notes constituting a concept [logic]

Thus, we can say that analysis is the identification of some object or concept through its breakdown into the basic elements that constitute it.

THE MEANING OF FUNCTION ANALYSIS

Function Analysis Analyzes the Actions of Products and Services

Goods and services have various actions. The aim of function analysis is to reveal each of these actions and thus identify the nature of the products and services to which they belong.

Function Analysis Analyzes the Purposes of Products and Services

All products and services are endowed by their creators with certain purposes. Function analysis reveals the intentions or purposes behind the creation of a product or service and thereby identifies the nature of that product or service.

Function Analysis Analyzes the Concepts of Products and Services

Although products and services exist as physical objects or systems, they are not created out of nothing. They are preceded by an idea — a concept — which is the basis of their creation. Function analysis identifies the nature of products and services by revealing these concepts.

THE FUNCTION ANALYSIS APPROACH

In function analysis, concepts are considered to be purposive actions involved in the creation of products and services. In terms of classical logic, the analytical approach in function analysis is to analyze the nature of things teleologically, that is, in terms of their purposes. Table 1-1 compares[1] a teleological

Table 1-1. A Comparison of Analytical Approaches

	Method of Analysis	Object of Analysis	Features of Analytical Method	Techniques of Analysis
Descriptive analysis	Descriptive analysis is an analytical method that answers the question, "What is it?"	Takes as its objects things and phenomena that we perceive, e.g., processes, tasks, quality characteristics	Analyses are performed differently by different people, but set analyses are possible when analytical methods are codified in rules.	• Process analysis • Task analysis • Motion analysis
Cause-and-effect analysis	Cause-and-effect analysis is an analytical method that answers the question, "Why is it so?"	Takes as its object the relationships between phenomena or events, e.g., the cause-and-effect relationship between defects and complaints	• Analyses vary according to situation, level of problem awareness, etc. • No analytical result is the theoretically "correct" answer • Is a practical analytical method	• Cause-and-effect diagrams • Fault tree analysis (FTA) • Failure mode effect analysis (FMEA)
Analysis in terms of purposes	Analysis in terms of purposes is an analytical method that answers the question, "Why is it so?"	Takes as its objects humans and human behavior, e.g., purposes, intentions, motivations, functions	• Lacks logical precision • Is a commonsense method of analysis	• Function family trees • Function analysis system technique (FAST)

approach (analysis in terms of purposes) to other analytical methods.

The function analysis approach is based on the third of the analytical methods shown in Table 1-1. We can now summarize this approach as follows:

- Function analysis, a method of analyzing in terms of purposes, allows you to analyze the achievement of previously set or recognized goals.
- Function analysis is a commonsense approach to understanding human beings and human behavior in terms of purposes, intentions, motivations, and functions. It cannot be called a scientific method.
- Function analysis does not have the logical precision of the natural sciences. This is not because of any inadequacy in the analytical method (or in logic) itself, but because the fact that the objects treated by that analytical method (or logic) are "soft" systems of human beings and human behavior.

FUNCTION ANALYSIS AND VALUE ENGINEERING

The Scope of Function Analysis

Function analysis analyzes — and identifies the nature of — concepts as purposive actions involved in the creation of products and services. The process consists of three procedural steps (Figure 1-1).

Understand the Object of Analysis

The use of function analysis requires an accurate and correct understanding of the object of analysis. Such an understanding, based on the collection of data, is a prerequisite to an accurate understanding of the concepts that form the purposive

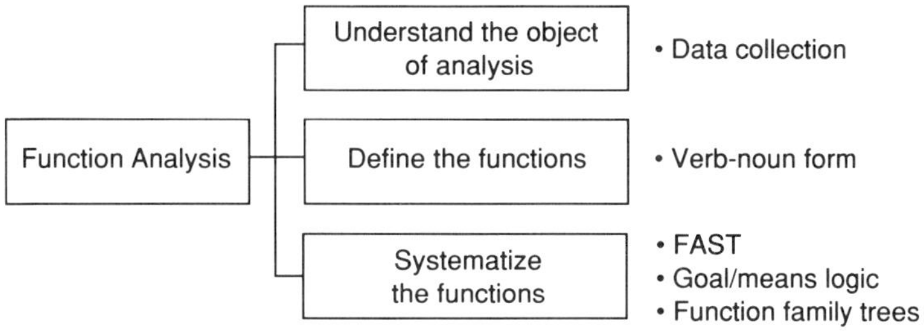

Figure 1-1. The Function Analysis Approach

actions of the object of analysis: What functions is the object of analysis supposed to perform? How do those functions relate to one another?

Define the Functions

Function definition lies at the heart of function analysis. There may be more than one function an object is supposed to perform. If there is a primary function, there may also be secondary functions. Function definition uses a verbal model (expressed in verb-noun form) to reveal each function. Since functions are by nature concepts, function analysis uses verbal models to define individual functions.

Systematize the Functions

Like function definition, function systematization occupies a central place in function analysis.

The various functions an object of analysis is supposed to perform do not exist in isolation; they possess specific relationships to one another. The function analysis system technique (FAST) and *goal/means logic* are methods for systematizing functions. Charts that systematize the relationships among functions are referred to as *FAST diagrams* or *function family trees.*

Function Analysis and the VE Job Plan

Function analysis is both the foundation of value engineering (VE) and its most characteristic analytical technique.

VE techniques can be summarized in terms of *VE job plans* (see Figure 1- 2).[2] The functions of products and services are analyzed and their value improved through VE job plans. Within VE job plans, the domain referred to as function analysis may designate either the basic function definition step and the function evaluation step, or it may designate only the basic function definition step. We will define function analysis according to the latter fashion: as the basic function definition step.

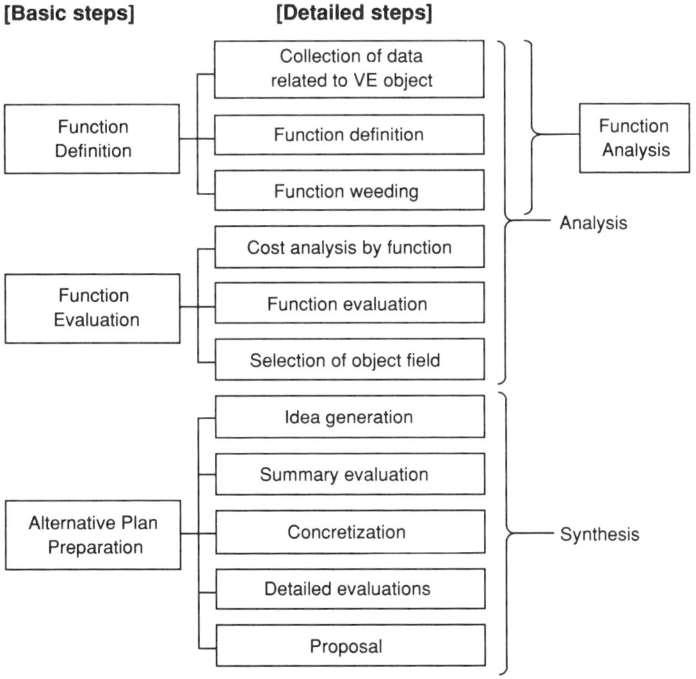

Figure 1-2. VE Job Plan

2
A History of Function Analysis

THE CREATION OF FUNCTION ANALYSIS BY LAWRENCE D. MILES

Function analysis did not come into existence and develop as an isolated entity. It was born, along with value analysis, as a basic value engineering technique and has continued to develop within the framework of VE.

When Lawrence D. Miles developed the technique called "value analysis" in 1947, he laid the foundation of modern VE. The following points concerning the identification of functions and definitions in VA appear in Miles's first book, *Techniques of Value Analysis and Engineering.* [1]

> Customer-use values and customer-esteem values translate themselves into functions as far as the designer is concerned. The functions of the product . . . cause it to perform its use and to provide the esteem values wanted by the customer.
>
> The use values require functions that cause the product to perform, while the esteem values require functions that cause the product to sell.

IDENTIFY THE FUNCTION PRIMARY/SECONDARY FUNCTIONS
TWO WORDS

> Understanding the desired function of a product or item or the expected performance of a service is the basis for determining value content.

Miles's statements emphasize that a clear grasp of intended function is the cornerstone of VA. The first of the three fundamental steps of VA is to "identify the function." Miles explains this as follows:

> Any useful product or service has a prime function. That function can usually be described by a two-word definition such as, provide light, pump water, indicate time, exclude dust, support handle.
>
> In addition, there may be secondary functions involved. A light source may be required to resist shock, a pump for domestic use to operate at a low noise level, a clock or watch to provide attractiveness. . . . [These] are secondary functions. . . . A basic ground rule in this connection is that we must accept the assumption that the device or part is in principle what is wanted to provide the function.

From the above, we see that expressing function as a verb and its direct object is the basis of VA or VE, and that this method is the starting point of function analysis within VA.

THE GROWTH OF FUNCTION ANALYSIS[2]

The Birth of FAST

Miles developed the function analysis method of expressing function as a verb and direct object. But there was no way in Miles's method to systematize multiple defined functions and identify a basic function among them. Charles W. Bytheway came up with the idea of applying several questions to individual functions in order to isolate the basic function from among many other functions. These questions carried an unexpected

IN IDENTIFYING PRIMARY VS. SECONDARY FUNCTIONS, FAST NOTES THE RELATIONSHIPS AMONG THEM

boon, for not only did they identify basic functions, they clarified the conceptual relationships among individual functions.

In 1965, Bytheway narrowed his tests down to nine key questions and announced the development of the function analysis system technique (FAST), a method for systematizing functions by applying the questions.

FAST refers to a method for systematizing functions; a chart made up of the results of FAST is called a FAST diagram. A FAST diagram is evidence of a completed function analysis. Bytheway explained the significance of FAST as follows: "The value of a completed FAST diagram is not so great compared to the value of the thought and creativity that went into making it." For Bytheway, in other words, FAST is something that stimulates and motivates creativity in the process of its application; it is therefore nothing less than a tool for creativity.

The Birth of the VE Function Chart

Arthur E. Mudge's VE function chart also contributed to the development of Miles's function analysis method of expressing function as verb and direct object.

In 1965, at nearly the same time that Bytheway's FAST appeared, Mudge published an essay entitled "How to Construct and Use a VE Function Chart,"[3] thus heralding a new method of function analysis.

Mudge explained that a VE function chart, which "clearly indicates unnecessary functions, overlapping functions and functions necessitating high costs," allows one to "evaluate one's own design objectively and realize lowered costs without damage to the essential role of the product." The VE function chart itself lists the essential function of the object product at the top and the functions of its parts below. Although there are procedures for constructing a VE function chart, there is no particular logic in the links among functions. We may therefore be justified in calling Mudge's VE function chart a method of systematizing functions aimed at lowering costs.

As we have seen, function analysis became a method for the systematic analysis of functions with the growth of Bytheway's FAST and Mudge's VE function chart — methods deriving from Miles's original function analysis technique of expressing functions with verb-noun pairs.

THE DEVELOPMENT OF FAST [4]

FAST for Planning and Control

The range of application of Bytheway's FAST method grew, extending into planning and communications.

D. P. Goss has promoted FAST as an effective tool for improving planning and resource allocation as well as for gaining external acceptance of plans. S. C. Fowler has called for the use of FAST as an effective communications tool. And W. F. Ruggles has praised the effectiveness of FAST as a vehicle for mutual understanding in planning.

The process of applying FAST and the resultant FAST diagrams are undoubtedly effective in the realms of planning and control. This use of FAST, moreover, has added the temporal notion of *when* to the *why* and *how* logic of FAST diagrams.

From FAST to FAST Diagrams

Bytheway sees the real significance of FAST as lying in its application. But he focuses his attention on the diagrams that result from applications of FAST. Ruggles proposes a fixed format for diagrams constructed by means of FAST and advocates drawing up diagrams in conformance with that format. Ruggles's version of FAST uses the logic of *why, how* and *when,* and he introduces the notions of superordinate functions, essential functions, secondary functions, permanent functions, simultaneous functions, design objectives, and problem domains.

T. J. Snodgrass also advocates a version of FAST that gives a fixed format to diagrams. Snodgrass's FAST diagram refers to superordinate functions as "tasks" and below them places essential functions and four supplementary functions. He designates these four supplementary functions as assuring convenience, reliability, satisfaction, and appeal, and writes that their purpose is to secure customer "acceptance" of the product.

The Ruggles and Snodgrass versions of FAST, with their patterned, formulaic diagrams, represent interpretations not seen in Bytheway's FAST. We can fairly say that FAST has changed course as its significance and emphasis have shifted from the application process to the diagrammed result.

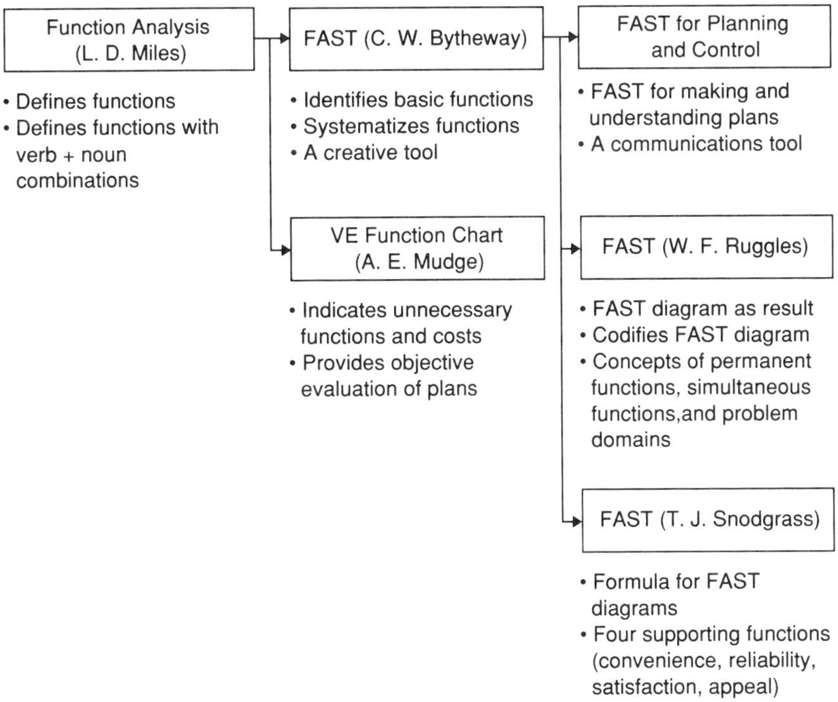

Figure 2-1. The Growth of Function Analysis in the United States

THE DEVELOPMENT OF FUNCTION ANALYSIS IN JAPAN

The Birth of Function Family Trees

Miles's function analysis method of defining functions by means of verbs and direct objects was introduced into Japan along with VE around 1955.

Between 1965 and 1975, Mudge's VE function chart appeared in volume 2 of the Japan VE Association's journal. In 1967, in volume 8 of *Value Engineering,* the late Professor Masatosi Tamai published an article entitled "How to Build Function Family Trees," based on Mudge's VE Function Charts and Bytheway's FAST. That same year, Professor Tamai authored the book, *Function Analysis,* in which he offered a critical analysis of Mudge's VE function charts, Bytheway's FAST, and function family trees.

Professor Tamai introduced the expression, "function family tree" in his 1967 article. At the same time, he cited six goals for the construction of function family trees. Table 2-1 compares these goals to those of Mudge and Bytheway.[5]

Table 2-1. Goals of Function Analysis

Bytheway's FAST	Mudge's VE Function Chart	Tamai's Function Family Tree
• Identifies basic function • Systematizes functions • A creative tool	• Indicates unnecessary functions and costs • Provides objective evaluation of plans	• Checks suitability of functions • Discovers neglected functions • Identifies function fields • Reveals high-order/low-order function relationships • Teaches concept for design to perform a certain function • Identifies basic functions

In contrast to Mudge and Bytheway, who stressed the application of VE function charts and FAST, Professor Tamai described the construction of a function family tree in technical terms, thereby emphasizing the method's technical aspects. He also defined FAST as a technique for analyzing functions and the function family tree as a diagram resulting from an analysis of functions. Thus, it can be said that, of the two function analysis techniques, FAST stresses the analytic *process* and function family trees emphasize analytic *results*.

Since its introduction in 1967, the function family tree has been used in Japan as a method of function analysis.

The Development of Function Family Trees

Of the three function analysis methods discussed by Professor Tamai — Mudge's function charts, Bytheway's FAST, and the function family tree — it was the technical diagram of results, the function family tree, that took root in Japan. The Japan Value Engineering Association published the report of its FAST study group as *A FAST Manual* in 1976, but despite comparatively widespread knowledge of FAST throughout the VE field, the technique itself was seldom used. This is borne out by a survey conducted in 1984 by Professor Masayoshi Tanaka of the Science University of Tokyo.[6] The survey summarizes results of questionnaires from 123 Japanese companies that are implementing value engineering. Table 2-2 shows the proportional use of function analysis according to the survey.

Table 2-2. Relative Use of Various Function Analysis Techniques

Technique	Percentage Use
FAST	26%
Function family trees	66%
Other	8%

Among function analysis techniques in Japan, function family trees used as a means of systematizing functions have taken root along with the method of defining functions with verbs and direct objects. In response to the expanded scope of VE application, the function family tree has developed in a variety of ways and is now used in function analysis for product improvement, product development, and software design and improvement. In addition, it shows signs of leaving VE and developing more broadly as a design and management method. (See Figure 2-2.)

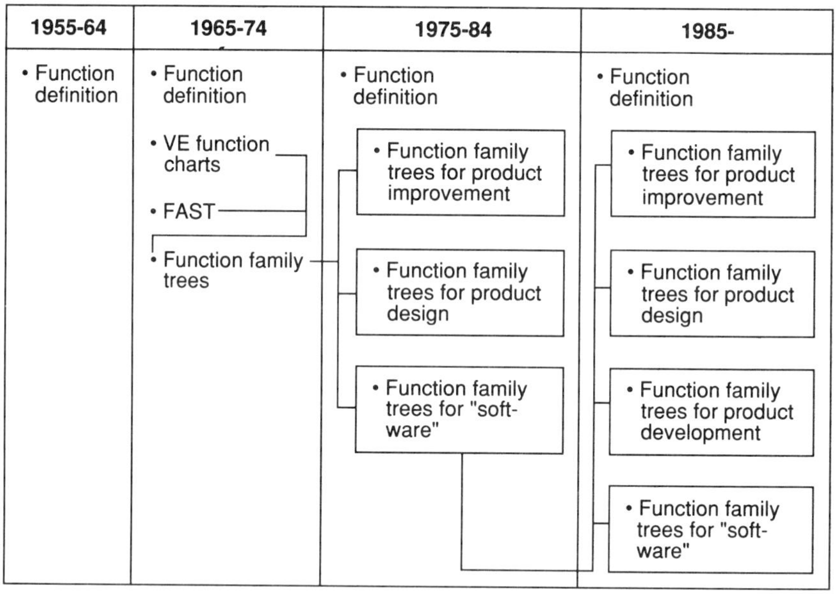

Figure 2-2. The Growth of Function Analysis in Japan

3

The Need for Function Analysis

PRODUCTS

Products are meant to be used by customers or "users." Companies need to provide low-cost products that conform to the users' purposes.

All companies experience numerous problems with the products they manufacture and supply to the market. In addition to technical problems involving quality, safety, ease of handling, service, and aesthetic design, they encounter economic problems relating to issues such as price and usage costs. A company's mission is to resolve these problems and to continue providing users with high-value products.

Any kind of "problem resolution" — even when unrelated to products — demands analysis of the problem. Problem resolution is composed fundamentally of two procedures: analysis and synthesis.

- *Analysis* — Isolate the problem by identifying the factors that generate it.
- *Synthesis* — Establish methods to eliminate the problem and draw up improvement plans.

Issues in Product Problem Analysis

Few suitable analytic techniques have in the past been available for resolving product-related problems. Indeed, there have even been attempts to resolve problems without really analyzing them. The methods used often fail to address a variety of issues relating to the analysis of product problems, such as quality and safety.

- Plans for improvement are considered without analyzing problems.
- Ad hoc analyses take place when problems occur.
- Problems are treated and analyzed individually.
- Problem analysis means analysis of the physical parameters governing product materials (such as tolerances and dimensions).
- Attempts are made to analyze the results of problems directly.

Function Analysis for Resolving Product-related Problems

These general issues related to the analysis of product problems are making their presence felt at more companies than ever before. Not only does the failure to resolve these issues make it impossible to solve product problems appropriately, but it also prevents companies from supplying users with high-value products. The reason for this is that conventional analytical methods have not resolved underlying problems. Rather, they have been temporary, partial, and based on appearances.

Products are, first and foremost, things to be used by customers. Companies should have the users' purposes and intentions in mind as they develop and design products, manufacture them, and supply them to the market. The occurrence of product problems means simply that the users' needs and company's intentions are not being fulfilled. When that happens, you have to analyze the problem by going back to those needs and intentions. Table 3-1 outlines the need for function analysis.

Table 3-1. Product Problem-solving and the Need for Function Analysis

Shortcomings of Current Problem-solving Approach	Function Analysis for Problem Solving
Focuses on improvement without analysis	Is accurate; focuses on function analysis
Relies on haphazard analyses	Is based on systematic function analysis of products
Relies on *ad hoc* analyses	Is based on systematic function analysis of products
Is based on analyses of physical product attributes and conditions	Is based on function analysis of product-use goals
Is based on analyses of results and appearances	Is based on function analysis of product's essence

Function analysis treats user needs as product functions and analyzes the problem from a functional perspective. Analyzing a product problem in terms of function reveals the fundamental nature of the problem and leads the way to a fundamental solution.

PRODUCT DESIGN

The Meaning of *Design*

The word *design* encompasses a vast number of concepts, but in general we can assign to it the following meaning: a scheme that gives concrete form to some goal or goals. Also, in manufacture or construction, design refers to the use of blueprints and other methods to clarify plans relating to cost, site, materials, and various aspects of structure.

Thus we see that design means planning the form and determining the content of some human-made object in accordance with the goals or purposes of the object to be made. In terms of human thought, moreover, design is a process that begins with an awareness of goals and then determines the content of products or systems by addressing those goals through repeated creation and trials. We can thus explain the word *design* as follows.[1]

- Design is an activity that recognizes the goals or purposes of products or systems.
- Design is an activity that shapes its objects — creates their forms — in accordance with the goals or purposes of those objects.
- Design is an activity that evaluates and determines the forms of its objects and makes their contents universally comprehensible.

The Product Design Process

Product design is an activity that identifies the purposes of the product that constitutes its object and then evaluates and determines the form of that product at the same time that it creates methods for achieving those purposes. It is also an activity that displays the resultant product form by means of drawings.

In contrast to the product manufacturing process, which manipulates materials physically, design creatively manipulates information on the basis of experience, knowledge, data, and so forth.

As a creative process of information manipulation, product design takes a myriad forms in different companies. Sometimes design is understood to mean simply the drawing of diagrams. Because product design is a creative process of information manipulation, it is sometimes difficult to standardize and establish it as an objective process. And even where a definable process has been established within a company, it is relatively common for work to proceed in a way that relies on the personal experience of the designer.

In general terms, however, product design is carried out according to the process shown in Figure 3-1.

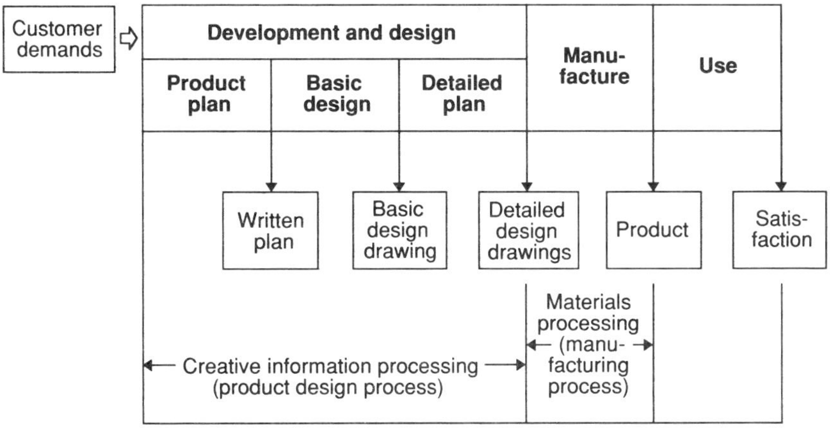

Figure 3-1. The Product Design Process

The product design process begins with a solid understanding of the customer's demands (the products' purposes or goals). The process then shapes the product as it moves from written plan to detailed drawings to evaluation and decision. The process, in other words, transforms abstract customer demands into detailed drawings of a product with a specific form. We may summarize the product design process as follows:

- The product design process transforms abstract customer demands into specific product drawings.
- The product design process is a process of function allocation that identifies product purposes — such as functions — and allocates them to a structural product.

- The product design process manipulates information creatively.
- The product design process is a decision-making process.

Issues in Product Design

The exercise of product design determines the basic components of product quality and cost. Product design, in other words, determines the fundamental components of product "goodness" or "badness." How is such product design actually carried out in companies and what sort of problems and issues arise? Table 3-2 gives an overall view of product design-related issues and measures.

Table 3-2. Product Design Problems and Approaches

Product Design Problems	Approaches (to Solutions)
Erroneous or nonexistent decision making	• Establish and perfect decision-making processes (i.e., product design processes), standardize • Collect data relentlessly • Express creativity • Exploit all design techniques • Guarantee schedules and man-hours
Inadequate information	• Set up and exploit information systems for design • Assemble complete information data bases • Collect information and data relentlessly
Insufficient ideas	• Generate a maximum of ideas • Keep extensive idea banks
Inadequate techniques	• Exploit all design techniques • (Function analysis, FMEA, standardization, etc.)
Scheduling or work-hour shortages	• Adhere to detailed daily plans • Commit labor needed

Function Analysis for Product Design

Product design identifies the purposes of the product (its object) and then evaluates and determines the form of that product while creating methods for achieving those purposes. It also displays the form of the resultant product by means of drawings. Finally, product design determines the basic components of product quality and cost.

Function analysis analyzes the functions of products or services as purposive actions. In other words, it analyzes the functions demanded of products and services when they are bought. To put it another way, function analysis analyzes the usage goals of products and services from a teleological point of view.

If product design is based on product goals, function analysis is no less than the basic analytical technique of product design. Table 3-3 shows specific results obtainable through the use of function analysis for product design.

Table 3-3. Effects of Function Analysis on Product Design

Product and Design Process Activities	Effects of Using Function Analysis
Recognition of object of design	A functional perspective provides a clear picture of the goals of the object of design
Creation of object of design	Designs develop smoothly on the basis of functions
Evaluation of object of design	A functional perspective provides clear criteria for design evaluation
Transferral of customer demands to drawings	Transferring customer demands to drawings becomes easier when those demands are seen as functions
Assignment of functions to product	Function analysis makes it clear which functions should be assigned to the product
Creative information processing	The processing of ideas and information can be creative when it is based on functions
Decision making	Decision making can proceed smoothly and surely when based on functions

PROCESS IMPROVEMENT AND PROCESS DESIGN

Definitions

The word *process* is generally used to mean the following:

- Manufacturing method, operation, stage, course, measure, method
- A set of action or changes in a special order

In contrast to such general definitions are the four goals established in industrial engineering (IE) — processing, inspection, transportation, and delay — as "unit processes." When these are linked together, they are referred to as a "process series." A process, therefore, is simply the course by which objects are transformed.

A process is sometimes referred to in a broad sense as a production system or a work system. Therefore, *process improvement (kaizen)* means improving the existing production system, and *process design* means planning and beginning the construction of a new production system. A production system as the object of process improvement or process design is shown in Figure 3-2.

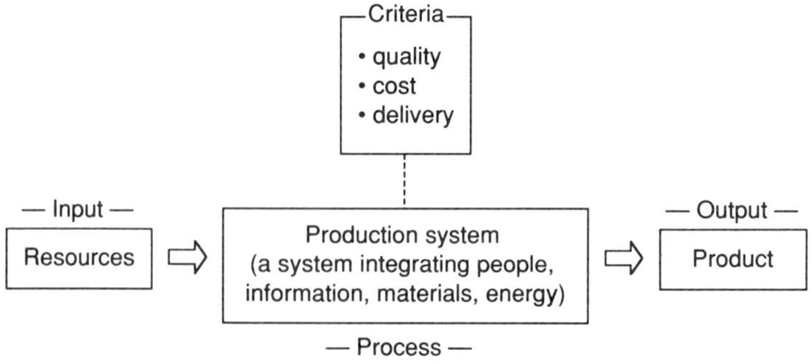

Figure 3-2. Production System

A production system designates a transformation process for generating output — such as products — from the input of various resources (people, information, materials, energy) under given conditions, such as quality, cost, and delivery schedule.

Process improvement, then, means making the production system better in terms of quality, cost, and delivery.

Process design means planning and constructing a new production system that guarantees quality, cost, and delivery.

Traditional Approaches to Process Improvement and Process Design

Process improvement and process design have traditionally been handled within the domain of industrial engineering. Advances in factory automation (FA) and the robotization of processes in recent years, moreover, have encouraged a trend in which process improvement and design issues are studied from the standpoint of automation. But there are problems with such traditional approaches.[2] Traditional approaches emphasize

- kaizen
- phenomena
- techniques
- means

Traditional approaches to process improvement and process design center on *kaizen*. This is particularly true in IE, where the approach is to analyze current processes and operations in detail, identify waste *(muda)*, and then eliminate it. Approaches of this sort work only with difficulty in situations involving the development and design of new processes. In addition, process kaizen does not permit radical improvement.

The second problem with traditional approaches to process improvement and process design is their focus on *observable phenomena*. A process is fundamentally a series of steps that physically transform material objects. When an examination of these steps begins with current processing methods, it precludes radical process improvement or process design. This is true of both industrial engineering and factory automation approaches. In FA especially, current processing methods — such as observable phenomena — are often simplistically transferred to machines.

The third problem with traditional approaches is their focus on *techniques.* Industrial engineering, for example, has established various techniques whose availability leads people to apply them to a range of problems and to hurry into reviewing processes without having accurately grasped the problem. Fundamental process improvement and design cannot take place under such conditions.

The fourth problem of traditional approaches is the focus on means. Equipment, tools, fixtures, and human work methods — these are all process means. In factory automation, problems are considered first from the point of view of robotizing hardware such as machines, tools, and fixtures. Industrial engineering focuses on human work. Yet no amount of examining such means can make radical process change possible.

Although traditional approaches include various techniques for dealing with process improvement and design, they are not necessarily systematic. No universal approach exists, moreover, for process design.

Function Analysis for Process Improvement and Process Design

As stated earlier, process improvement means making new and better production systems in terms of quality, cost, and delivery. Process design means planning and beginning to construct new production systems in terms of quality, cost, and delivery. And function analysis analyzes functions as the purposive actions of products and services. Applied to processes, this means that it analyzes purposive actions intended to transform physical objects — such as functions for transformation.

Using function analysis for process improvement and process design eliminates problems inherent in conventional approaches and can lead to results such as those shown in Table 3-4.

Table 3-4. Effects of Function Analysis on Process Improvement and Design

Emphases of Traditional Approaches	Effects of Using Function Analysis
Kaizen	A functional perspective (a) clearly indicates which transformation functions a process should perform and (b) makes a design approach to production systems possible.
Phenomena	Designing and improving processes on the basis of knowing what transformation functions should be performed makes it possible to build production systems that break free of the phenomenological pull of traditional methods.
Techniques	Production systems that address specific problems or goals can be designed and improved by examining processes in terms of transformation functions they need to carry out.
Means	Production systems can be designed and improved based on specific goals of processes by examining processes in terms of transformation functions they need to carry out.

SERVICE AND INDIRECT MANAGEMENT ACTIVITIES

Characteristics of Service Operations

Despite the widespread use of expressions such as "service" and "service operations," the terms have no clear definitions.

Things intended for consumers are generally referred to as "commodities," and commodities may be further broken down into "goods" and "services." The former are generally referred to as "merchandise" or "products." Table 3-5 shows features of services that distinguish them from merchandise or products.

Table 3-5. Distinguishing Characteristics of Services

	Merchandise and Products	Services
Form	Tangible	Intangible
Storage	Can be stocked. Place and time of production and use are different.	Cannot be stocked. Place and time of production and use are the same.
Type of Industry	Primary, secondary industries, including manufacturing industries.	Tertiary industries, including retail sales, finance, transport, and telecommunications.

One characteristic of services is that they are intangible — they cannot be picked up and handled. This means that quality and cost may not be as clear as they are in the case of physical products. Similarly, the concepts of manufacture and design are vague in the service field.

Another characteristic is that services cannot be stored; their creation and use are simultaneous. Consider, for example, riding on a train or being waited on in a restaurant. With services, in other words, there is no time lag between production and consumption. This feature sharply distinguishes services from products, in which there *is* a time lag.

A third characteristic has to do with type of industry. Industries that produce merchandise or products make physical items, whether by natural or technological means. Industries that render services either do not handle physical items (as in the case of finance or transport communications) or — even if they do — produce something peripheral to physical items (as in the case of retail sales).

Characteristics of Indirect Management Activities

In a broad sense, indirect management activities possess the same characteristics as service activities. As in the latter domain, indirect management activities have in the past seen a multitude of approaches to improvement and design.

The absence of a decisive feature is itself a distinguishing characteristic of indirect management activities and has contributed to the difficulty of improvement and design in the field. In Table 3-6, we show its characteristics by contrasting it with shop floor tasks in manufacturing industries.

Table 3-6. Characteristics of Indirect Management Activities

Work Type / Work Characteristics	Manufacturing Work	Indirect Management Work
Work content	Physical work: handling physical objects (materials, products, equipment, etc.)	Mental work: handling information and knowledge
Work repetition	Mechanical, fixed, repetitive	Little repetition, changes according to circumstances
Work visibility	State and progress of work easy to see	State and progress of work difficult to see
Work evaluation	Easy to evaluate by quality, costs, schedules	Work evaluation measures vague, evaluation difficult
Work improvement	Improvement easy for above reasons	Improvement difficult for above reasons

The characteristics of indirect management activities shown in Table 3-6 can also be seen as features of services.

Service tasks involve more mental work than manufacturing tasks do because, even though services may require considerable labor, the work deals with people and information.

In addition, the fact that service tasks tend to involve the handling of situations rather than the performance of repetitive operations makes them more difficult to perceive than manufacturing work. Other features of service work are that criteria for evaluation are vague and improvement is difficult.[3]

Function Analysis for Service and Indirect Management Activities

Compared with product fabrication and manufacturing work, service and indirect management activities tend to be vague. Like the former types of work, however, service and indirect business activities exist to perform certain functions. Function analysis makes it possible to improve and design service and indirect management activities rationally.

Table 3-7 shows examples of specific results obtained in these fields through the use of function analysis.

To summarize, function analysis (1) objectively identifies work constituents from a function perspective and (2) permits radical improvement and design based on functions.

QUALITY CONTROL

Definitions of Function and Quality

The use of the word *function* is not nearly so widespread as that of the word *quality*. Quality in the context of quality control has been described as follows:

- that which matches buyers' demands (Japan Industrial Standards)
- degree of usability (Deming)

Table 3-7. Effects of Function Analysis on Services and Indirect Management Activities

Characteristics of Services and Indirect Management Activities	Effects of Function Analysis
Work is not visible, tangible.	The content of work becomes visible, thus simplifying improvements and design.
Storage is impossible; production and use occur simultaneously.	Waste can be eliminated from production (i.e., the provision of a service or carrying out of a support activity) at the place of use by improving and designing work at the planning stage.
Evaluation of work is difficult because evaluation measures are vague.	Clear function-based evaluations become possible.
Planning and improvement are difficult.	Radical designs and improvements become possible.

- a reinterpretation or social reevaluation of utility value determined by preproduction estimates (Hashimoto, Ninzō, et al.)
- fitness for use (Juran)

These descriptions use different words to say the same thing: that the quality of a product designates its fitness in terms of usability or use.[4] As Figure 3-3 shows, however, there is a wide gap between underlying quality and quality as actually managed in companies.

Function designates the purposive actions of services and products — concepts that cannot be captured in terms of forms or colors. Figure 3-4 sketches the relationship between the meaning of function and the meaning of quality.

Issues in Quality Control

The Japan Industrial Standards (JIS) publication *JIS Z 8101* (1981) defines quality control as follows:

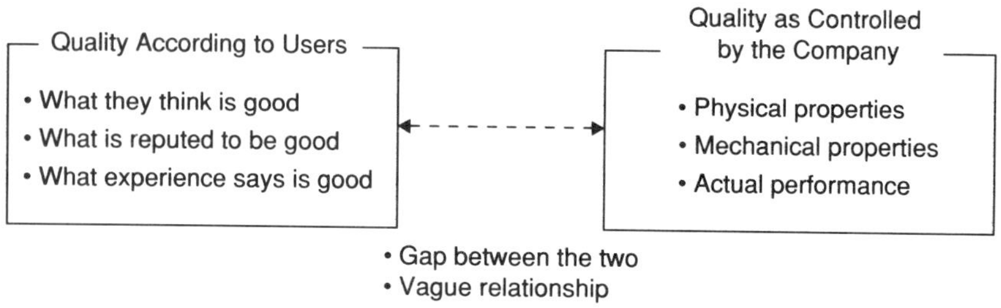

Figure 3-3. The Quality Gap Between Users and the Company

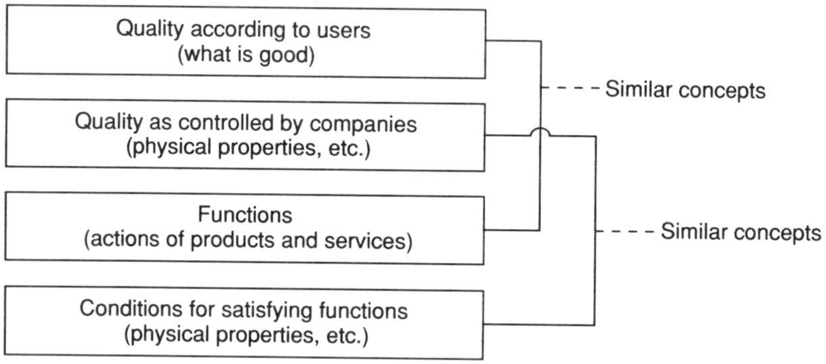

Figure 3-4. Function versus Quality

> Quality control is the system of all measures taken for the purpose of economically manufacturing goods and services whose quality matches the demands of buyers. Modern quality control makes use of statistical means and so is often characterized as statistical quality control (SQC).

Defined in the above manner, quality control has in recent years been addressing itself to all aspects of company activities in the form of total quality control (TQC). Examining what really goes on in companies reveals various actual quality control problems that need to be resolved. The following problems are representative:

- It is not always clear how to define quality as demanded by customers.
- Quality control emphasizes the product manufacturing stage.
- Quality control focuses on the physical characteristics of products.

The first problem is that existing methods for identifying customer demands relating to quality — techniques including quality charts and quality deployment — are not always adequate for grasping the nature of quality. The usual quality control approach is to measure and then control variability in the properties of parts or products. But what the customer wants are product functions or ease of use determined by those properties. The term quality includes the concept of product functions or ease of use, but quality control techniques do not provide adequate means for identifying and analyzing those features.

The second problem is that quality control is still applied largely at the manufacturing stage. Producing goods whose quality satisfies customers' demands requires making quality explicit in — and integral to — the product planning and design stages. This shortcoming has its roots in the first problem — that of inadequate techniques for analyzing product functions and ease of use.

The third problem is that quality control applies principally to physical products. The concept of quality ought to apply to all goods and services created by human beings. Traditional quality control, however, has focused on methods for analyzing variability in the properties of quantifiable parts and products. This has made it difficult to define and analyze the quality of services and other nonquantifiable products. The natural result has been a kind of quality control that tends to concern itself with quantifiable aspects of parts and products.

Using Function Analysis for Quality Control

The underlying purpose of quality control is to identify and guarantee the kind of quality demanded by customers. But quality control methods don't adequately tell you what sort of quality it is that customers want. As a result, quality control as currently practiced tends to be limited and to emphasize the physical properties of products in the manufacturing stage. Function analysis analyzes functions as the purposive actions of goods and services. Customers and users buy products and services because they want the functions those products and services provide. A quality control system in tune with those functions accurately targets quality as conceived by users and as demanded by customers.

Table 3-8. Effects of Using Function Analysis

Shortcomings of Quality Control (QC)	Effects of Using Function Analysis
Vague grasp of quality demanded by customer	• Identifies quality demanded by customer in terms of functions • Identifies all aspects of quality demanded by customer • Clarifies the meaning of quality demanded by customer
Primary emphasis on manufacturing stage	• Makes it possible to apply QC at the planning and design stages • Opens the way to "life cycle" QC linking, planning, design, and manufacturing
Primary emphasis on physical properties	• Makes QC possible at the level of required functional properties • Makes QC based on functional properties possible in fields other than manufacturing

The use of function analysis makes possible true quality control — quality control centered on users and customers. Table 3-8 shows some specific effects of function analysis on conventional quality control.

In addition to the effects shown in the tables, function analysis allows for creative, goal-oriented quality control. Traditional quality control methods have taken a data-oriented control approach, but function analysis is teleological in nature. Function analysis opens the way to a quality control approach based on customer demands.

4
Techniques of Function Analysis

FUNCTION DEFINITION

In the context of function definition, *function* means the purposeful action of a product (in service industries, administration, office work, processes, operations, and so forth). *Definition* refers to the process by which the content of the function is made clear. Thus, the basic meaning of the word define is as follows:

- *Define:* Determine the boundaries of, delimit, make clear, identify the meaning of.
- *Identify:* Recognize as identical, distinguish one thing from another; confirm, verify.
- *Clarify:* Make clear, limpid, lucid.

Thus, *to define* means to delimit, distinguish, and identify something. But in terms of function definition, this doesn't go far enough. We need to take our well-defined entity, (a) make it universally understandable, and (b) give it objective expression by naming it.

Function definition means defining functions in the terms just explained. The Japanese term for definition is seldom used because it is not the social custom in Japan to define things or to act on the basis of contracts. The lack of such conventions, rooted in the country's geographical and historical influences, results from the Japanese tendency to keep things vague and to settle matters ambiguously.

The Significance of Function Definition

Physical Properties and Function Properties

Manufactured products are physical objects. But those who use the products want more of them than their physical properties; they want the *function* that the product performs. (Thus, users of a washing machine want the removal of dirt from clothing; users of a refrigerator want food preservation and ice making.) They also look at aspects of the product's appearance, such as its form, color, or general attractiveness. Function analysis refers to the first class of functions as "use functions" and to the latter as "esteem functions."

Manufactured products, then, have physical — or technical — properties that relate to their physical structures. At the same time, they possess function properties demanded by customers or users.

Most manufacturing companies put their chief efforts into physical properties of the products, such as structure, materials, and dimensions. The overriding concern of customers and users, however, is function properties — properties to which physical properties are only a means.

Function definition reveals each function as it focuses on the function properties that make up the goals of products. (See Figure 4-1.) The thorough examination of a product's function properties via function analysis opens the way to product improvement or development based on the perspectives of users.

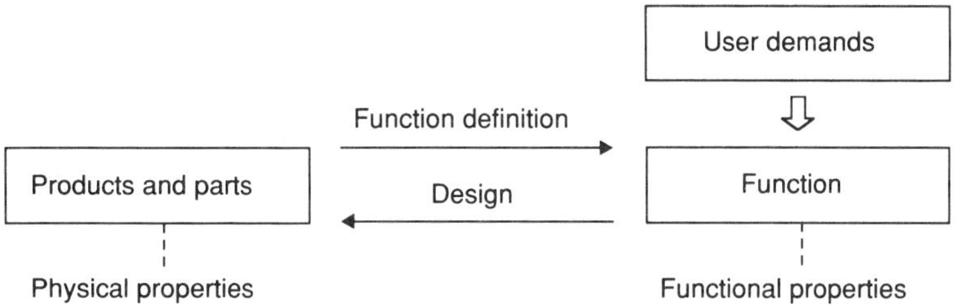

Figure 4-1. Converting User Requirements into Physical Products

Design and Function

Design involves perceiving the purposes of a design object, creating the means to achieve those purposes, and then evaluating those design objects. Seen as a process, design consists of converting user requirements into drawings — of assigning required functions to physical products, as in Figure 4-1.

What we are referring to as design objects and user requirements are, in fact, functions. Function definition gives us a clear understanding of both the design objects and the customer and user demands. At the same time, it enables you to proceed with design predicated on customers and users.

At the most basic level, design engineers are people who give concrete form to the demands of customers and users. Function definition makes it possible for the work of design to proceed rationally.

Goals of Function Definition

Function definition is a basic function analysis technique with four goals:[1]

- to identify the functions of the product or service to be designed or improved

- to evaluate the relationships between costs and the functions of the product or service to be designed or improved
- to create valuable products or services
- to evaluate alternative products or services

Identifying Functions

Improving or designing a product or service begins with a clear understanding — based on function definition — of the functions demanded for the particular product or service. Identifying functions also engenders a shared appreciation of the problem by every company employee with a stake in improvement and design. This eliminates miscommunication among sales, design, manufacturing, and other departments.

Evaluating the Cost/Function Relationship

The ultimate goal of function analysis is to improve and design value-bearing products and services so that required functions are performed at a low cost. For this to happen, you must identify each function through function definition so that you can then evaluate the relationships between functions and costs.

Creating Products and Services

Customers and users require the functions of products and services. Function analysis clearly identifies those functions, on whose basis you can then generate ideas and construct alternative plans. In other words, function analysis gives rise to value-bearing products and services from the perspective of the customer and without the constraint of past assumptions.

Evaluating Products and Services

Products and services exist to carry out functions demanded of them by customers and users. By identifying each function needing to be performed, function definition provides a clear basis for customer-oriented evaluations of whether improved or designed products or services satisfactorily carry out their functions.

How to Define Functions

How to Express Functions

Defining the function of a given object involves asking the question *What is its action (= function)?* and then expressing the answer in terms of a verb and its direct object. In function analysis, in other words, function is expressed through a verbal model that combines a verb and a noun (see Table 4- 1).

Table 4-1. Verbal Model of Function Analysis

Object of Function Definition	Verb	Noun
(Product or Service)	(Function)	
clock	indicate	time
	decorate	wrist
fountain pen	release	ink
	store	ink
fuse	cut off	excess current
binder	hold together	documents
	classify	documents
receipt	confirm	amount paid
travel report	convey	travel activities
production meeting	set	a production plan
cigarette	relieve	stress
tie	decorate	the chest

Expressing a function with a verb and a noun satisfies the goals of definition by allowing the function to be stated simply. Other reasons can be cited for using the verb-noun model, as well. Specifically, this model:

- makes it possible to identify each function clearly

- makes functions understandable to anyone
- makes it easier to come up with ideas

Keeping the following points in mind, we should express functions using:

- simple language anyone can understand
- specific, well-defined terms that preclude misunderstanding
- words independent of the object whose function is being expressed

How to Use the Verbal Model

A function is expressed as a verb and a noun: "Do something." Keep in mind the following points relating to the usage of each term:

Verbs. The verb is the predicate of a function expression — it is the verb, we might say, that actually expresses the function. Verbs can be classified as either transitive (that is, taking direct objects) or intransitive (that is, not taking direct objects). Transitive verbs are used for expressing functions because a function represents some action applied externally (Table 4-2).

Table 4-2. Verbal Model, Example 1

Light Bulb Function	Production Meeting Function	
• Provide light • Glow	• Set production plan • Production plan is set	(preferred) (not preferred)

Nouns. Nouns in the formula are usually direct objects. If we express the function of a light bulb as "provide light," *light* is the direct object of *provide.* It is best to steer clear of direct objects with a structural function (that is, "anchor motor" or "hold cover") and to avoid using the names of parts or products. The reason for this is that using specific part or product names robs the function of its universal meaning (Table 4-3).

Table 4-3. Verbal Model, Example 2

Cigarette Function	Switch Function	
• Relieve stress • Emit smoke	• Open/close circuit • Open/close switch	(preferred) (not preferred)

Abstract Function Expressions

A function defined in terms of a verb and a noun becomes both a topic and a catchphrase for idea generation and the creation of valuable alternative approaches. This means that, while a function expression must, of course, express the function accurately, it should also allow an extensive range of ideas when its verb-noun combination is used to generate new thinking.

In other words, function expressions should be abstract and not limited to specific means. (See Table 4-4.)

Table 4-4. Examples of Abstract Function Expressions

Object of Function Definition	Preferred Expression	Expression not Preferred
• Cigarette lighter • Drill press • Binder • Briefcase • Guest register	• Produce heat • Make holes • Hold together documents • Hold together carried items • Announce visitors	• Produce flame • Cut holes • Bind together documents • Contain carried items • Record visitors

Expressing a function by means of a verbal model is tantamount to abstracting that function, simply because words are abstract concepts. The scope of ideas is not enlarged, however, if an abstract concept captured by a function expression is constrained by the narrowness of that expression.

Expressing the function of a cigarette lighter as "provide flame" restricts new ideas to those that presuppose "flame." Expressing the function as "provide heat," on the other hand, opens the way to a variety of ideas not based on flames. The underlying function of a lighter, after all, is not to provide flame but to provide the heat needed to ignite tobacco.

In the example of a guest register, expressing the function as "record the arrival of guests" leads to discussions of ideas for recording. Moreover, the expression *record* does no more than express in words the current state of the guest register. Here, if the underlying function is to inform someone of a guest's arrival, "record" should be replaced by "inform of" or "announce." This would allow for ideas that might even eliminate the register itself.

As the previous examples illustrate, special attention needs to be paid to both the verb and the noun when making a function expression more abstract. This process of abstraction should be applied not only to the individual words but also to the meaning of the verb-noun expression.

Although in making a function definition more abstract you expand the meaning of the function, you also need to clarify the bounds of the function's meaning by making the function expression appropriately specific. It is vital that expressions be both suitably abstract and specific.

Easily Quantifiable Function Expressions

A function defined by a verb and a noun serves as both a prerequisite to the evaluation of ideas and alternatives and as a standard. A function expression should, of course, express function accurately. It should also be easily quantifiable in order to facilitate the evaluation of ideas and alternatives. You do this by making it possible to identify — and set criteria for — the degree to which the function is performed. (See Table 4-5.)

Table 4-5. Examples of Easily Quantifiable Function Expressions

Object of Function Definition	Preferred Expression	Expression not Preferred
• Fuse • Watch • Electrical cord • Receipt	• Cut off excess current • Indicate time • Deliver current • Confirm amount of payment	• Cut off electricity • Indicate hour • Deliver electricity • Confirm payment

In the case of a fuse, expressing function as "cut off electricity" makes it difficult to determine how much electricity to cut off — that is, to set numerical criteria for satisfaction of the function. This is because the word *electricity* is hard to convert into a scientific, numerical value. If we change "electricity" to "excess current," however, we can convert to a numerical value and therefore easily set numerical criteria for function performance. In other words, a criterion of function performance can be set such that the fuse should "cut off current at *n* amperes." This also opens the way to clear evaluations of ideas and alternatives for satisfying the function. Evaluation becomes simply a matter of whether a given device can cut off excess current or not. The same argument can be applied to an electrical cord.

To use another example, the action or function of a receipt is obviously that of proving that money was paid, and thus it can be expressed as "confirm payment." But the word *payment* provides no criterion for judging the extent to which the function has been performed, because it does not indicate how much payment needs to be verified. It becomes easier to set a standard when we specify what sum needs to be verified through an expression such as "confirm amount paid."

Thus, we see the importance — especially with respect to nouns — of making function expressions easily quantifiable. But in some instances it is more beneficial to use verbs that can be easily converted into numerical values. Keep in mind that function expressions need to be both abstract and easily quantifiable.

At the same time, since not all functions can be quantified and not all words can be converted easily into numerical values, easy quantification should not be made an absolute prerequisite for a function expression.

Setting Criteria for Functions

A function can generally be characterized by the degree to which it performs as intended. Moreover, functions are ordinarily fulfilled under certain conditions. Degrees of fulfillment correspond to performance or reliability; and performance conditions, necessitated by human or environmental factors, correspond to maintainability, operability, and safety. These are called "criteria" for functions.

Types of criteria. We can determine criteria for a function by subjecting that function to the so-called 5W1H questions: *who, what, when, where, why,* and *how much.* Figure 4-2 gives an example of these questions applied to a light bulb's "provide light" function.

Function

Why
"provide light"

Criteria

• Who —	ease of operability and absence of hazards, e.g., excessive heat in providing light (safety, operability)
• What —	weight (*n* g) and dimensions (*n* cm) of bulb providing light (material conditions)
• When —	maintenance of function of providing light *n* hours or more (reliability, durability)
• Where —	resistance to temperature (*n* degrees) and humidity (*n* %) of place used to provide light (conditions of use environment)
• How Much —	brightness (*n* lux) of light provided (performance)

Figure 4-2. Asking the 5W1H Questions

not 'why' but 'how much'

Sometimes additional factors, such as methods, patents, or utilities, constrain the performance of a function.

Criteria and quality characteristics. Criteria for functions include quality control factors such as demand quality, demand characteristics, and substitute characteristics.

In quality control, quality is managed by assigning and then controlling numerical values for fluctuations in what function analysis refers to as criteria. Function analysis itself identifies conditions that constrain functions and then uses them as standards for evaluating ideas or alternatives. In other words, function analysis establishes and evaluates what quality control calls quality by identifying it in terms of constraining conditions and then making it a prerequisite to the performance of functions.

Broad and Narrow Senses of "Function"

In function analysis, the meaning of function may either be confined to a verb-noun expression or it may include criteria. (See Figure 4-3.)

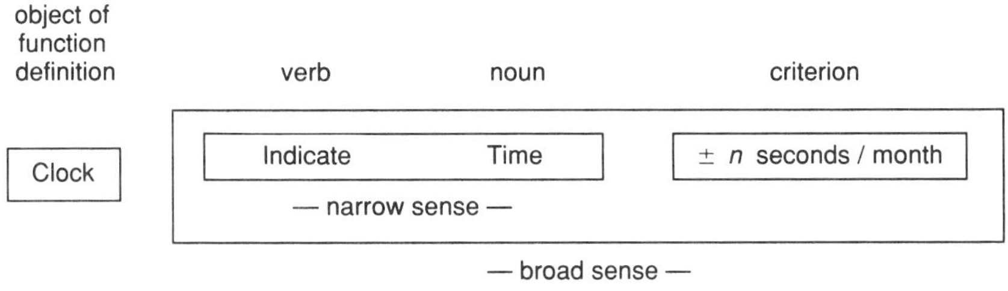

Figure 4-3. *Function* in the Narrow Sense and in the Broad Sense

Used in its narrow sense, function is a topic or catchphrase in the process of idea generation. The broad meaning of function includes the degree to which the function performs as intended, and it is used in the evaluation of ideas or alternatives.

In the case of a clock, the narrow meaning of function would be "indicate time," and this would be used as a topic when ideas are generated. But the indicated time may be inaccurate; that is, it may not correspond to the actual time. Clocks are constrained by margins of error of *n* seconds per month, and this margin of error becomes a criterion by which function-related ideas or alternatives are evaluated.

The concept of setting criteria. Criteria for functions are dependent on the demands of users and determined on the basis of various specifications and ratings.

Not all criteria are needed for all functions. Because function analysis is based on the users' perspective, the only criteria that need to be established are those based on specifications or ratings received from the users. Moreover, excessively rigid criteria restrict the means available to satisfy the functions. Liberal criteria allow a broader range of methods to fulfill functions. At the same time, however, strict criteria facilitate evaluation of ideas and alternatives, while liberal criteria make such evaluation difficult.

When setting criteria, it is important to understand the concept and nature of constraints.

Classes of Functions

Functions can be classified from a variety of perspectives. Figure 4-4 shows types of functions and their relationships.

Use Functions and Esteem Functions

The term *use function* is applied to functions involving the purposes or goals for which products are used. Nearly all functions relating to the internal mechanisms of products, their parts, or materials fall into the category of use functions.

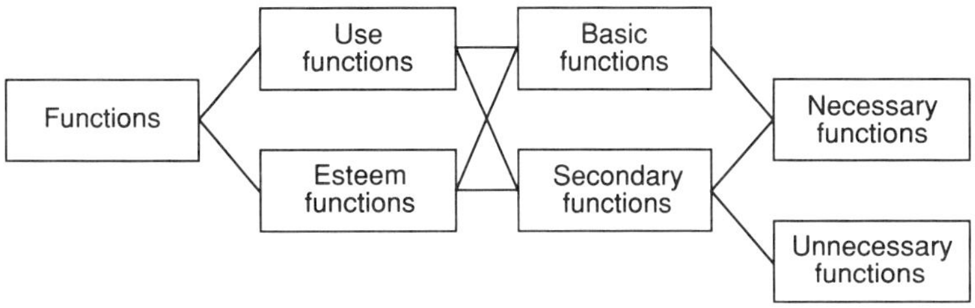

Figure 4-4. Function Types

Esteem functions relate to the forms or colors of products; the expression designates functions intended to give sensory satisfaction to the product user. Ordinarily, the field referred to as artistic or decorative design addresses itself to these esteem functions.

In recent years, esteem functions have taken prominence over use functions for items such as household electronics.

Basic Functions and Secondary Functions

Basic functions are functions that give the object of function definition its reason for existence. Basic functions will therefore vary according to the object and scope of function definition.

Secondary functions are all other functions that assist in the realization of basic functions. Secondary functions therefore also vary according to the object and scope of function definition.

Basic and secondary function classifications both vary according to the object and scope of function definition and are relative to one another.

Necessary Functions and Unnecessary Functions

Necessary functions are those demanded by the customer or linked to those demands. Unnecessary functions are neither demanded by nor linked to demands by the customer, and they can be eliminated by changes in the current design. They may also be functions that improvements in the current design would make unnecessary.

FUNCTION SYSTEMATIZATION

Systematizing functions means drawing relationships between the individual functions that need to be performed by the object of analysis and then showing those relationships on a chart. Functions are related to one another by ends/means logic and how/why logic. A function family tree (Figure 4-5) is a chart showing relationships based on the former kind of logic, while a FAST diagram (Figure 4-6) relates functions based on the latter type of logic.

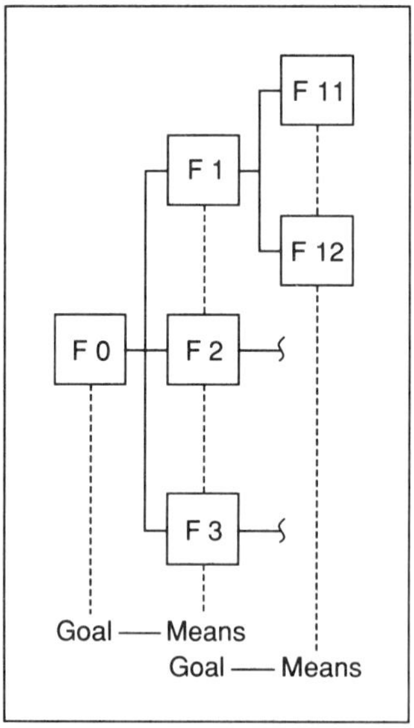

Figure 4-5. Function Family Tree

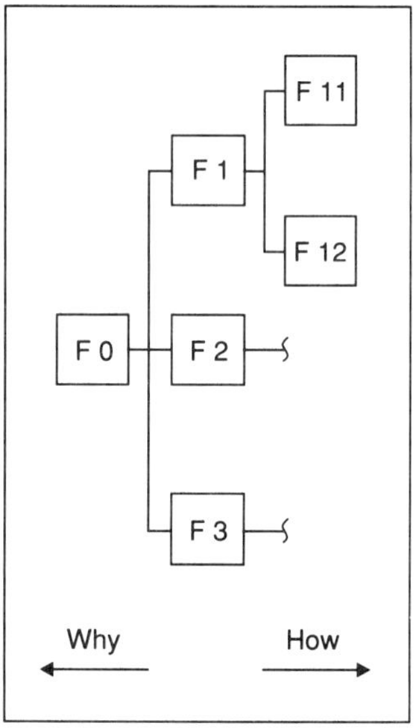

Figure 4-6. FAST Diagram

There are a variety of ways to go about explaining things.[2] Principal types of explanation include descriptive explanation, explanation via cause-and-effect relationship, and explanation via goals. Function analysis systematizes functions by means of this last type (explanation via goals), that is, through the logic of ends and means.

Ends/means logic is a method by which things are analyzed as entities so that some previously established or recognized goal(s) can be achieved. The analysis of people and human behaviors in terms of their goals, purposes, motivations, and functions is an example of ends/means logic. It follows that ends/means logic cannot have the logical rigor of the natural sciences. To summarize its chief characteristics, ends/means logic:

- takes human purposes and behaviors as its objects
- involves continuous links
- provides abstract ends and concrete means
- provides criteria for controlling human actions
- reflects human value orientations

The Significance of Function Systematization

Understanding the Object of Analysis

We generally understand products and services on the basis of their perceivable aspects. We understand a physical product, for example, by looking at its structure and form. We understand a service by observing the business activity or behavior by which it is performed. Yet products and services exist basically to perform some function or functions. We can easily understand the essence of products or services by systematizing the functions they need to perform. We do this by constructing function family trees and FAST diagrams.

From a technical standpoint, systematizing functions and constructing function family trees and FAST diagrams reveal the design concepts of products and services and makes those concepts universally understandable.

Stimulating Creativity

Systematizing functions and constructing function family trees and FAST diagrams are processes of tracking down those functions that constitute goals (such as "goal functions") and of using the question *Why* to pursue high-order functions.

The process involves thinking about goal functions — the sort of thinking that points to possibilities for improvement in current products and services. It may also lead to alternative methods for performing the functions. The process stimulates the creativity in all of us, in other words, and sparks a variety of new ideas. Charles Bytheway, the inventor of FAST, says that the greatest significance of the method lies in the stimulus that its application gives to creativity.

Improving Communication

People involved with the object of analysis are usually the ones who construct function family trees and FAST diagrams. The process of constructing these charts improves communication among team members by fostering a shared understanding of both functions and the links among functions. Mutual communication improves, moreover, when the function family tree and FAST diagram are shown to supervisors and managers with broader responsibilities for the object of analysis.

In other words, the process of constructing function family trees and FAST diagrams improves communication within organizational units involved with the objects of analysis.

Objectifying Function Relationships (Design Concepts)

Functions and their relationships are concepts. In the case of a tangible product, for example, the product's design concept originated in the mind of the designer. The same is true of a service.

Constructing a function family tree or FAST diagram objectifies a product's or service's design concept and makes it possible to represent design ideas externally.

Goals of Function Systematization

Like function definition, function systematization is a fundamental function analysis technique. It has three goals:

- to identify the high-order functions that the product or service to be designed needs to perform
- to identify the links among functions of the product or service to be designed or improved
- to create alternatives to the product or service to be designed or improved

Identifying High-order Functions

A number of functions are assigned to products and services. Some of these functions are partial and some need to be carried out by the product or service as an overall system.

Function systematization reveals which functions are partial and which need to be performed by the system as a whole. In other words, it clearly identifies the high-order functions that the product or service must satisfy as an overall system.

Identifying what functions need to be performed by the system is vital when you are designing or improving a product or service.

Identifying Links among Functions

The functions of products and services are interrelated in a variety of ways. Some functions exist to carry out other functions, and some functions cannot be satisfied without certain other functions. In other words, functions that products and services have to perform are variously interrelated and do not exist in isolation from one another.

Systematizing functions reveals the relationships and links among functions and makes it possible to represent those links objectively. It also brings to light the thinking behind the design.

Constructing Alternatives

When functions have been systematized in a function family tree or FAST diagram, that tree or diagram will show both the design concept and the ideal image, that is, how the product or service should look. You can then design a higher-value product or service by investigating ways to carry out needed functions based on this ideal image. Function family trees and FAST diagrams, in other words, permit the construction of high-value alternatives.

In addition, constructing function family trees and FAST diagrams helps you devise alternatives by systematically revealing needed functions — functions whose satisfaction is prerequisite to any improvement or design.

Function Family Trees

Function ends/means relationships. In the example of a pocket flashlight[3], the *function* of the open/close circuit is to "send current," and its *goal* is to "provide light." (See Figure 4-7.) But the *means* to the "provide light" goal consists of no less than five functions: "convert electricity to light," "send current," and so on. This is how a function family tree systematizes functions according to their ends/means relationships. Generally speaking, there are multiple means functions corresponding to a single goal function.

Function independent/subordinate relationships. The functions in the second row from the left of the chart ("provide light," "house part," and so on) are linked by independent/subordinate relationships rather than by ends/means relationships. In addition, they are all means for the "illuminate surroundings" function and are therefore subordinate to that function.

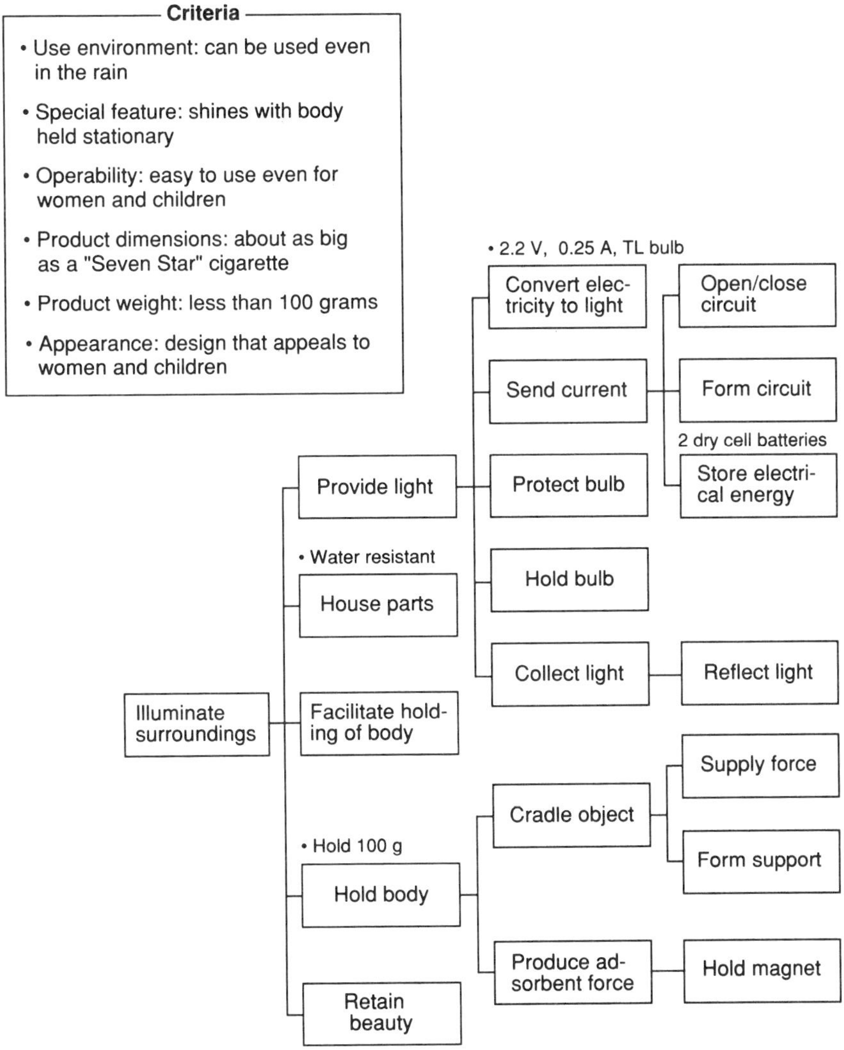

Figure 4-7. Function Family Tree for a Pocket Flashlight

High-order/low-order relationships. In the function family tree, functions on the left are referred to as "high-order functions" and functions on the right are called "low-order functions." The designations *high-order* and *low-order* are relative and in no way absolute.

Basic functions and secondary functions. The function at the extreme left of a function family tree is referred to as the "basic function" of the object of analysis. This basic function varies according to the scope of the system being analyzed. All functions to the right of the basic function are "secondary functions." In the flashlight example, the basic function is to "illuminate surroundings," and all other functions are secondary. Once again, the distinction between basic and secondary functions is relative rather than absolute.

Function levels and function fields. Functions at the extreme left of a function family tree are called "first-level functions." Moving column by column from left to right, they are succeeded by second-and third-level functions.

A group of low-order functions for which a given function is the ultimate end is referred to as a "function field." Function analysis generally takes second-level functions to be goal functions and their subordinate groups to be function fields. The flashlight example has five function fields.

Functions and criteria. Criteria that constrain functions are entered above the functions they constrain. In the flashlight example, there are six criteria — ranging from use environment to aesthetics — that constrain the "provide light" function of the overall system. There are, moreover, further criteria that constrain individual second-order functions.

Functions and criteria in a function family tree. In a system of functions, functions that (a) ensure the performance of high-order functions, or (b) are needed to satisfy conditions for high-order functions are placed below high-order functions. (See Figure 4-8.)

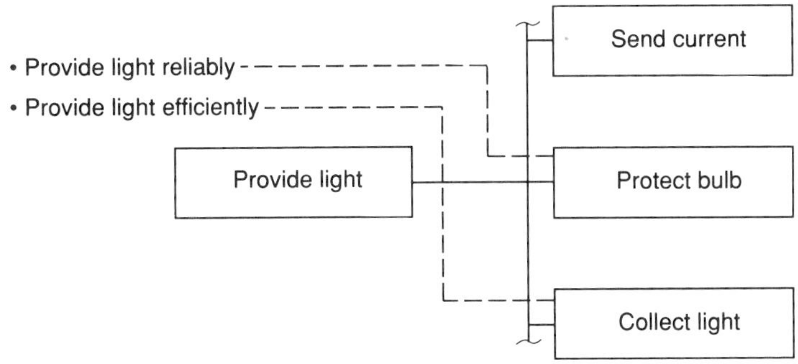

Figure 4-8. Functions and Criteria in a Function Family Tree

It follows that some functions are necessitated by conditions controlling high-order functions. Functions necessitated by such conditions vary with the particular conditions involved and the conditions themselves can be defined as functions.

The steps for constructing a function family tree are shown in Figure 4-9.

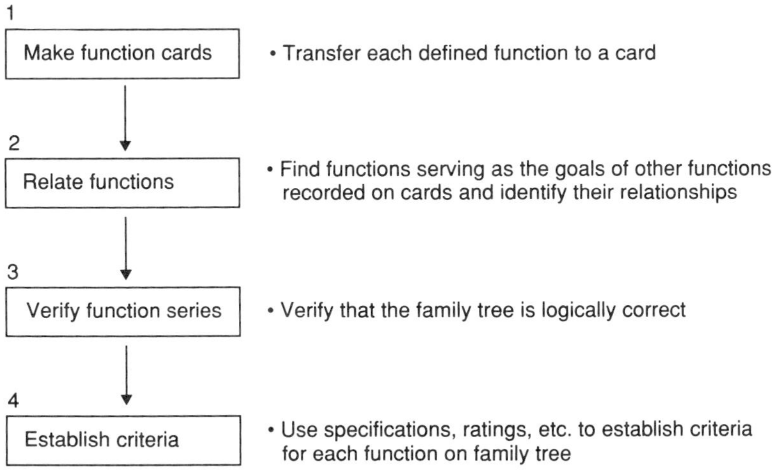

Figure 4-9. Steps for Constructing a Function Family Tree

The Bytheway Version of FAST — Logic [4]

See the example of the pocket flashlight in Figure 4-10.

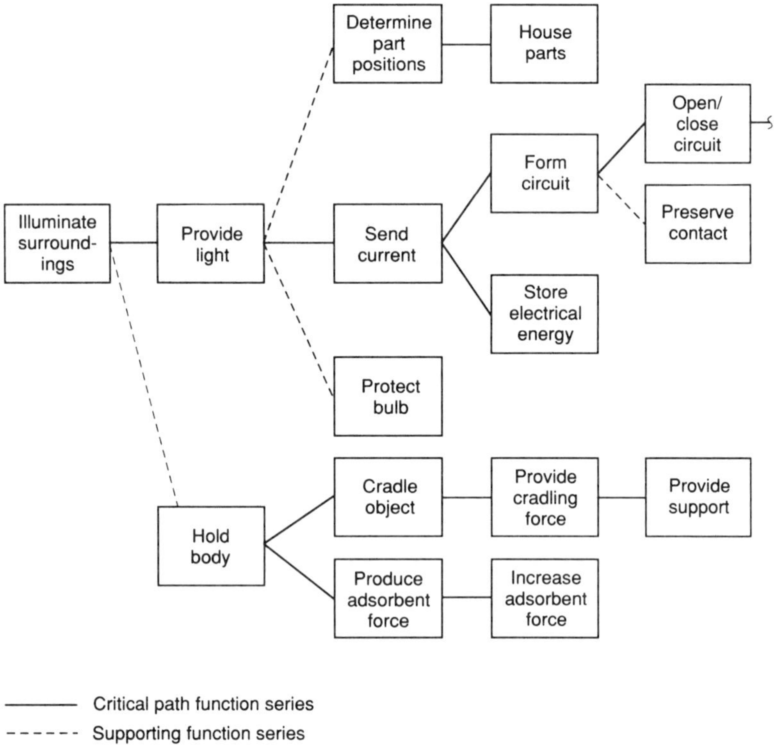

Figure 4-10. Bytheway's FAST Logic: A Pocket Flashlight

High-order function questions. Ask the following three questions of each defined function:

1. What do you really want to do in performing this function?
2. What high-order function is the reason for performing this function?
3. What is the function reason for having to perform this function?

The answers to these questions — along with the functions that prompted them — are treated as a "function group." The answers to the questions, moreover, become high-order functions. (See Figure 4-11.)

Function Group

- Defined function .... "provide support"
- Answer to question 1 .... "provide cradling force"
- Answer to question 2 .... "cradle object"
- Answer to question 3 .... "hold body"

Figure 4-11. Example of "Provide Support" Function for a Flashlight

Logic of determining the basic function. This is a protocol for determining which function of a function group is the basic function. You make the decision by comparing the functions with one another and asking the following question: *Would the product or service have to fill the other function if it didn't have to fill this one?*

If the answer to this question is yes, the other function is the basic function. (See Figure 4-12.)

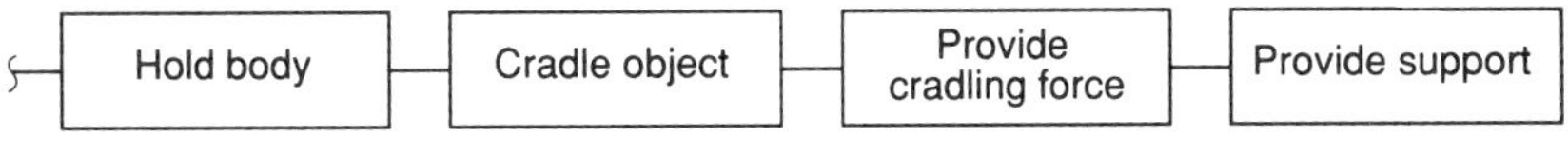

Figure 4-12. "Provide Support" Function Series for a Flashlight

Critical path function series. The following question is used to reveal the critical path series: *How is the function actually carried out?*

We begin with the highest-order function of the object of analysis and, step by step, move lower in accordance with the critical path. The answers to this repeated question must be specific. (See Figure 4-13.)

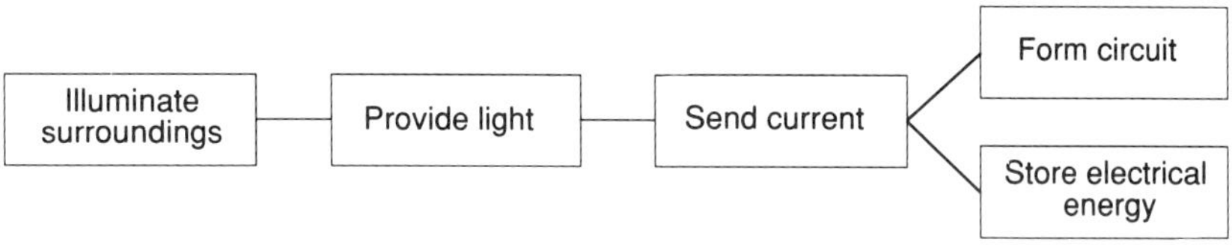

Figure 4-13. Example of Critical Path Function Series for a Flashlight

Basic function. The highest function of an object of analysis is called the basic function. The basic function generates all other functions of the same object.

Higher-order function. This term designates a function more important than the functions constituting the object of analysis. None of the functions constituting the object of analysis would be necessary unless the higher-order function were needed. Moreover, choosing an alternative method for performing the higher-order function may eliminate the need for the functions constituting the object of analysis.

Functions and criteria. A function is something expressed by a verb and a noun. Criteria are expressed by formulas, rules, or principles, and are used only for the purposes of evaluating or testing quality.

The Bytheway Version of FAST — Procedures

See Figure 4-14.

Mudge's VE Function Diagram — Logic[5]

See the example of the pocket flashlight in Figure 4-15.

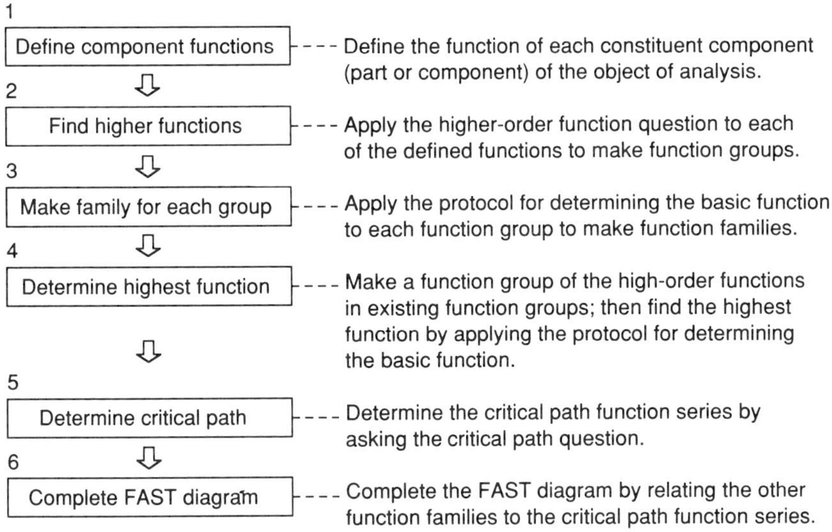

Figure 4-14. Bytheway's FAST Procedures

Assembly's basic function. When an entire assembly (object of analysis) is taken to be a black box, the basic function is captured in a verb-noun pair expressing the difference between the black box's input and output.

The basic function, in other words, is the action that exchanges input for output.

All assembly functions. This refers to all functions — including the basic function — that must be performed by the assembly, or object of analysis. In a VE function diagram, these functions are arrayed on the second row, below the basic function. In the flashlight example, "provide light," "send current," open/close circuit," "house part," and "hold body" are all assembly functions.

Part functions. This refers to functions that should be performed by the parts making up the overall assembly. In a VE

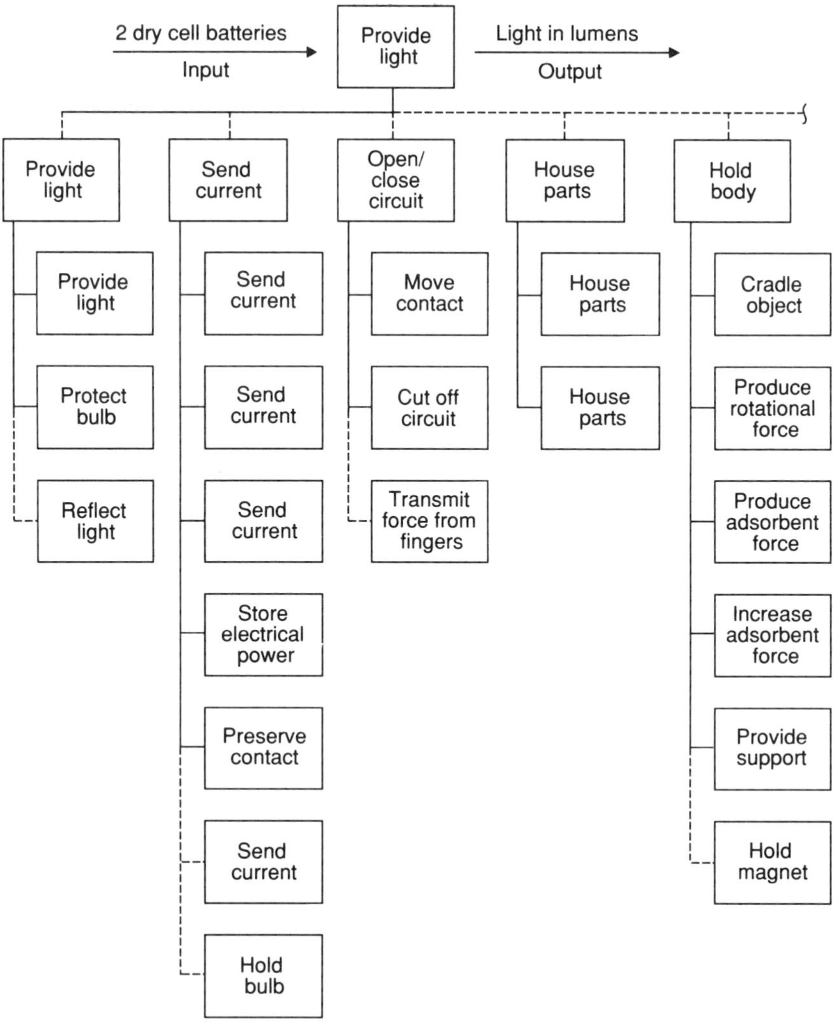

Figure 4-15. Mudge's VE Function Diagram

function diagram, these functions are placed below all functions believed to relate to all assembly functions.

Function level. Function level means the classification of parts and assemblies into basic and secondary functions. Based on this classification, the VE function diagram marks basic functions with solid lines and secondary function with dotted lines.

Steps for constructing a VE function diagram. See Figure 4-16.

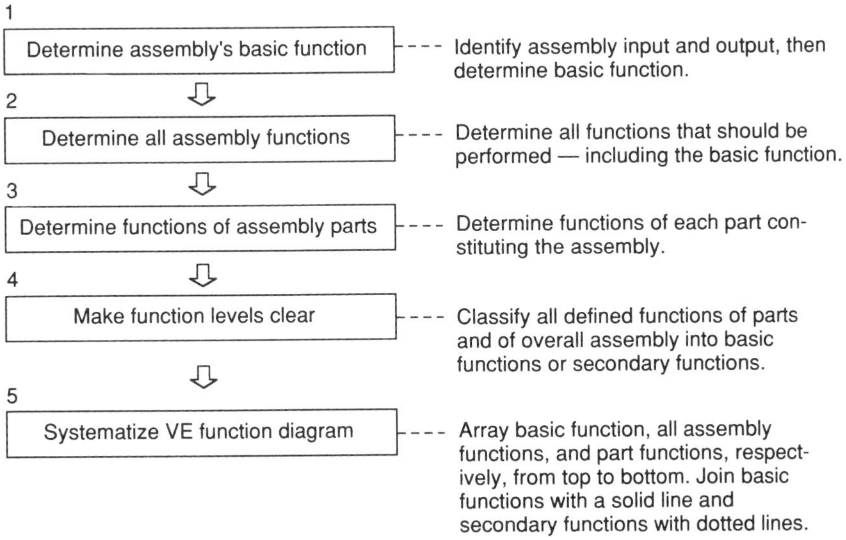

Figure 4-16. Steps for Constructing a VE Function Diagram

The Ruggles Version of FAST — Logic[6]

See the example of the pocket flashlight in Figure 4-17.

Problem domain. Draw vertical dotted lines to the left and right of the functions under analysis. The area between the two lines becomes the domain of functions addressed by improvement and design.

High-order function. This is the name given to the function immediately to the left of the vertical dotted line at the left of the problem domain. The high-order function is not an object

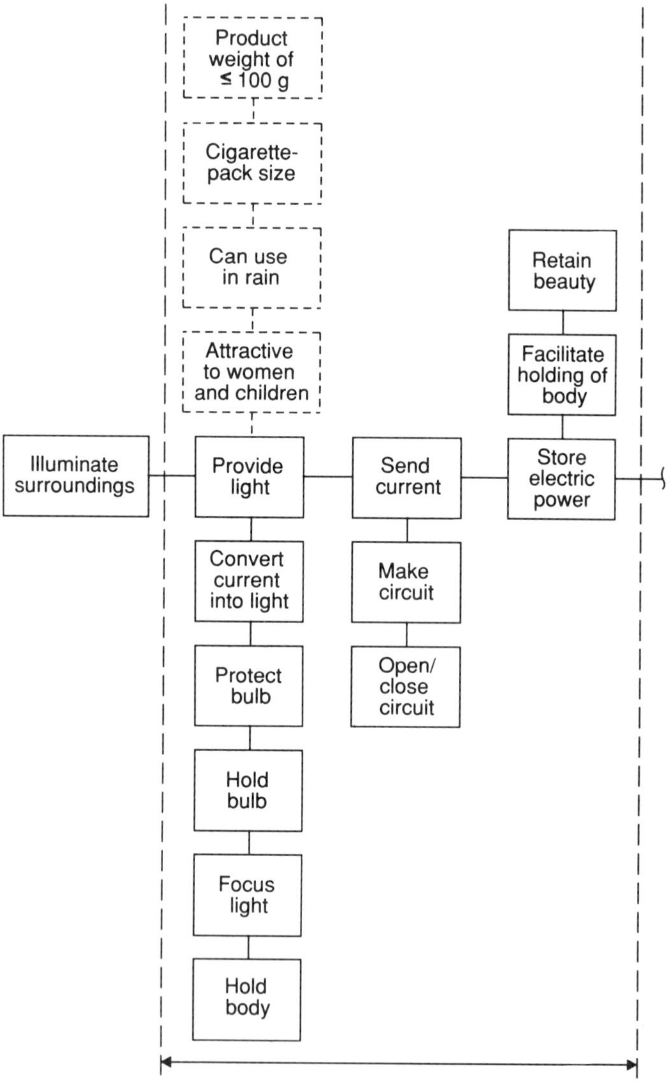

Figure 4-17. The Ruggles Version of FAST

or target of improvement or design. In the flashlight example, "illuminate surroundings" is the high-order function.

Basic function. The basic function, shown immediately to the right of the dotted line at the left of the problem domain, constitutes the ultimate goal of improvement or design activities. The basic function in the flashlight example is "provide light."

Secondary functions. To the right of the basic function are secondary functions — functions linked in terms of the critical path to *why* and *how* logic. For the flashlight, "send current" and "store electrical energy" are secondary functions.

Critical path functions. These are the functions joined by the *why* and *how* logic of the high-order function, basic function, and secondary functions. For the critical path, therefore, these are the only three kinds of functions that exist. Critical path functions are connected on the diagram by a solid line.

Simultaneous functions and permanent functions. Simultaneous functions are functions that have to be performed either because of or at the same time as functions along the critical path. They are placed below critical path functions on the diagram. Permanent functions are functions that always need to be performed. These functions are placed above the critical path functions.

Design objectives. Specifications and criteria involving improvement or design of the object of analysis are treated as design objectives and enclosed in dotted lines above the basic function.

The Snodgrass Version of FAST — Logic[7]

See the example of the pocket flashlight in Figure 4-18.

Task. In this diagram, the function furthest to the right is called the "task," and it tells what the assembly (object of

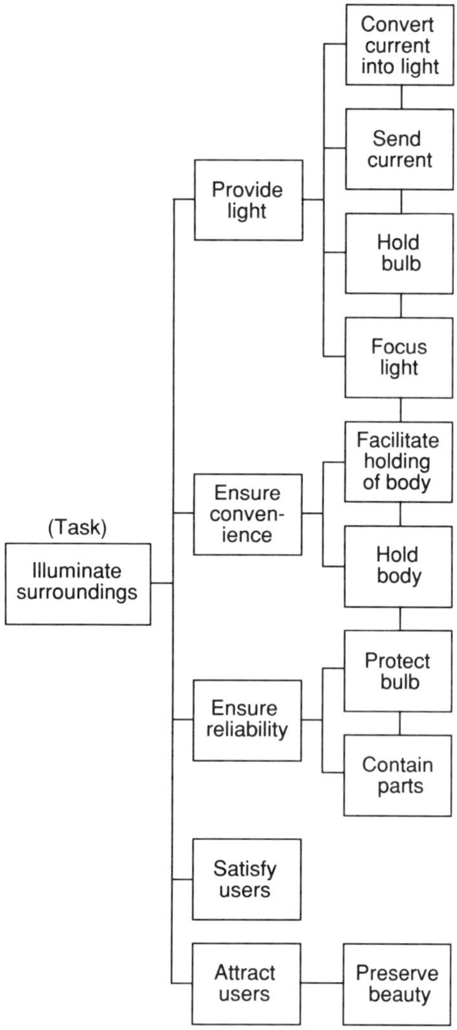

Figure 4-18. The Snodgrass Version of FAST

analysis) as a whole will do. The name *task* is used because it is more understandable to people than the term *function*.

Different assemblies have different tasks, but each assembly has only one. In the flashlight example, the "illuminate surroundings" function is the task.

Basic functions. Basic functions are located under the task and address the question of *how*. They are "essential functions," and most products will have three, four, or five of them. For the flashlight, "provide light" is a basic function.

Supporting functions. There are four supporting functions that secure a product's reception: (1) "ensure convenience," (2) "ensure reliability," (3) "satisfy users," and (4) "attract users." (See Table 4-6.)

Table 4-6. Four Supporting Functions

Supporting Functions	Description	Examples
Ensure convenience	Functions for making the product easier to use, more versatile	• Explain product • Explain operation
Ensure reliability	Functions aimed at product safety, reliability, maintainability	• Preserve safety • Prevent deterioration
Satisfy users	Functions to guarantee user satisfaction with the product	• Reduce noise
Attract users	Functions to make the product appeal to people	• Improve looks • Improve form

Primary and secondary functions. Basic functions and supporting functions are called "primary functions," and all functions below them are referred to as "secondary functions."

FAST for Planning[8]

The logical structure of FAST is shown in the example of an efficiency management system (see Figure 4-19). The procedure for FAST is outlined in Figure 4-20.

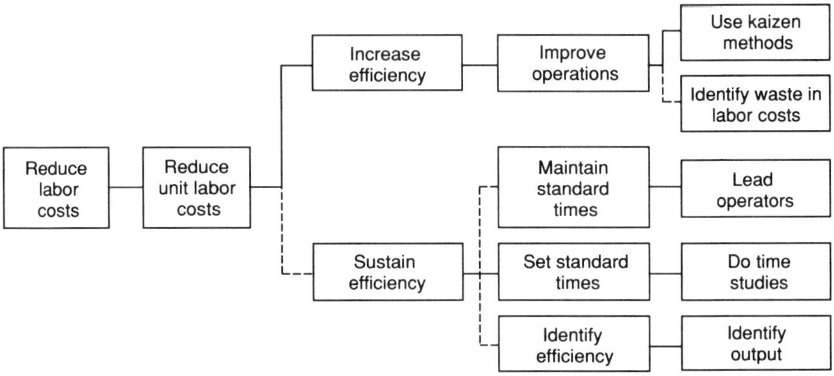

Figure 4-19. FAST for Planning

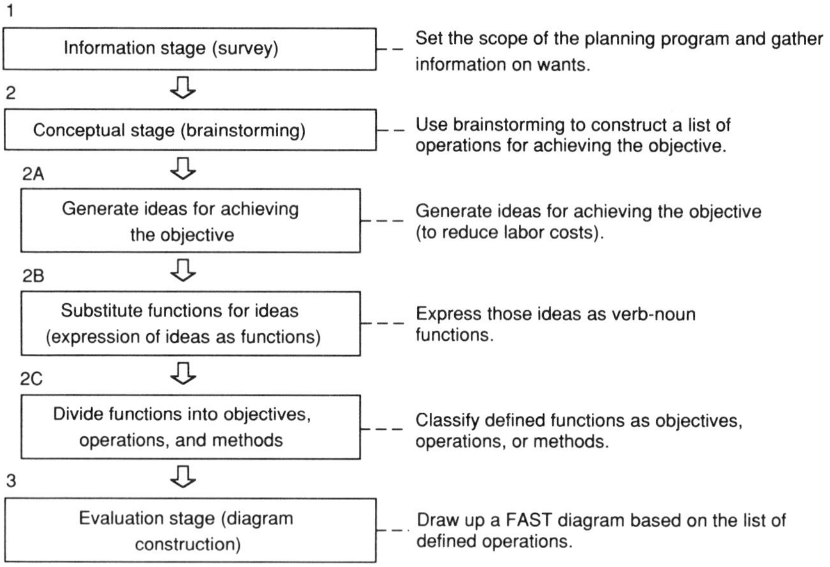

- Objective = the benefit the program is trying to achieve
- Operation = task (unit of work) that contributes to the objective
- Method = type of task for completing an operation

Figure 4-20. FAST Procedures

5
What Can Function Analysis Analyze?

FUNCTION ANALYSIS OBJECT SYSTEMS

What Is a System?

Function analysis is a technique for analyzing various systems, improving them, and designing new ones.

The type of system to which function analysis can be applied may be defined as follows:[1]

- two or more constituent components
- ordered interactions and functions among its constituent components
- specific overall goals and functions

A system is composed of two or more elements. A pin holding together the bricks or machinery of a building, for example, is a single element and not a system. Systems are ordinarily composed of tens or hundreds or thousands of elements and are by nature conglomerations of multiple constituent components.

Ordered mutual interactions and functions exist among the constituent components of a system. The components of a system are not haphazard; each possesses a set function. In addition, the individual functions of constituent components interact with one another to carry out other functions.

In an analog clock, for example, the hands have the function of indicating position. The internal mechanism has the function of turning a spindle in accordance with the actual time. The joint actions of the hands and the internal mechanism carry out the function of indicating the time.

A system, then, is something in which linkages and mutual interactions among components carry out other new functions.

A system refers to something with specific overall goals and functions. Systems do more than just exist. They have overall goals and functions they must perform. Indeed, systems exist *because of* their goals and functions.

A transportation system, for example, has the goals and functions of conveying people as rapidly, as safely, and as economically as possible. Similarly, a school system has the goals and functions of educating people.

Types of Systems

Static systems. A static system is an entity composed of multiple constituent components, whose object system is static. Bookcases, tables, streets, and tunnels fall into this category. Static systems to which function analysis might be applied to include products or parts of products, pieces of equipment, and tools or fixtures.

Dynamic systems. A dynamic system is a system in which some sort of energy input results in specific, predetermined motions. Automobiles and most machines belong in this category.

Products or parts of products, and machines or mechanisms are dynamic systems to which function analysis might be applied.

Control systems. A control system is a dynamic system with control functions — a system that maintains conditions within a prescribed range. Air-conditioner temperature control systems are examples of tangible control systems. Intangible control systems include the manufacturing processes and operations involved in production systems, as well as production control, inventory control, quality control, cost control, and other "control" systems.

Control systems for which function analysis might be used include tangible products and equipment, as well as all intangible management systems and the manufacturing processes and operations serving as production systems.

Response systems. A response system collects information from the external environment and then behaves in ways that respond to that environment. Response systems to which function analysis can apply include company management and strategic systems other than systems of administrative control, as well as various activities and systems relating to service.

Company systems. This term designates all company and public organization systems, from static systems to response systems. Organizational and service systems are examples of company systems to which function analysis can be applied.

Function Analysis Object Systems

Function analysis takes any system as its object. In other words, function analysis can be applied to all entities that qualify as systems and that have functions to perform (see Figure 5-1).

PRODUCT SYSTEMS

Functions

Products are intended to be used by customers (users). The functions of a product system therefore must all be demand-related functions based on the requirements of the customers.

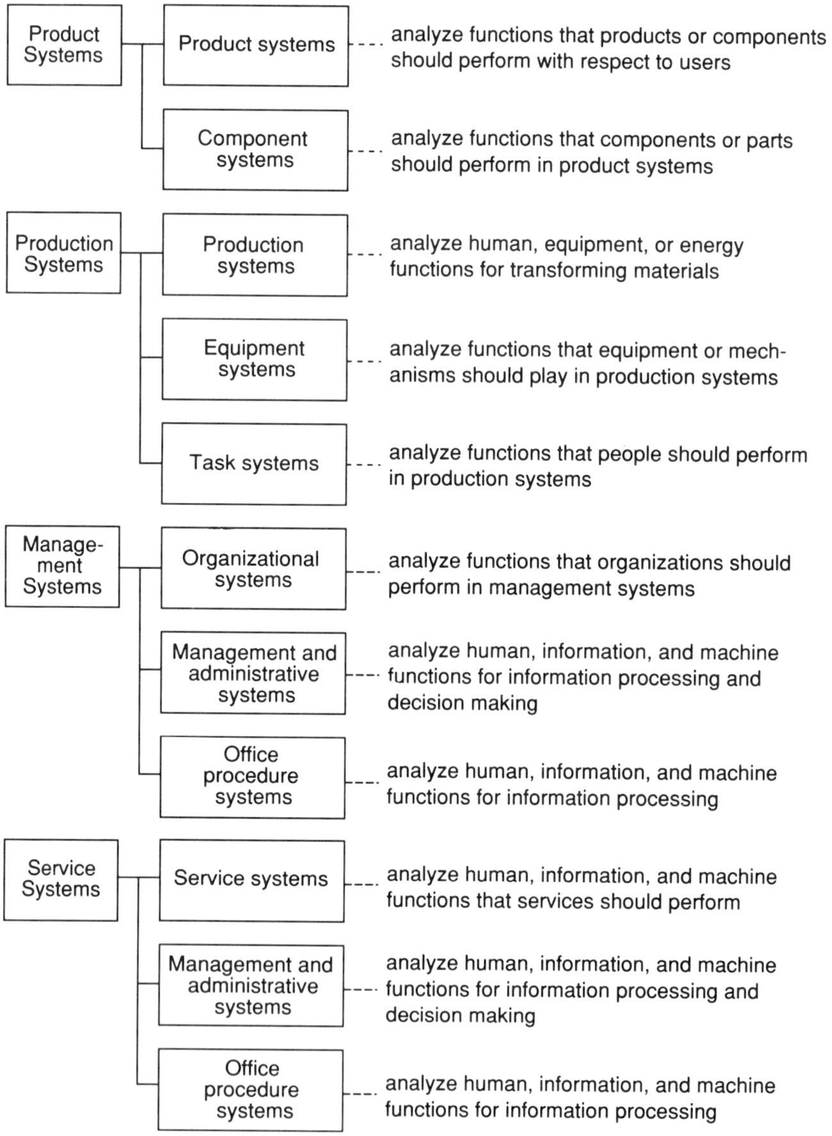

Figure 5-1. Job Analysis Object System

A product system is composed of parts and components, and the physical interrelationships among those objects carry out the functions.

Ease of Application

The functions of product systems are easy to analyze because they involve visible things (parts, components, products).

They are easy to analyze for another reason as well. Because product systems are complex entities composed of static, dynamic, and control systems, their behavior is stable and they carry out specific functions.

Modes of Application

When function analysis is applied to product systems, it can be applied either to the product as a whole or to one of its component systems (such as a subassembly). (See Figure 5-2.)

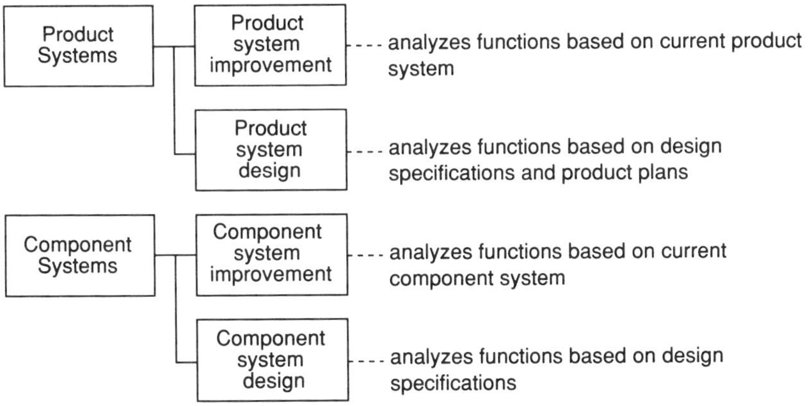

Figure 5-2. Modes of Application for Product Systems

Whether function analysis is applied to the product system as a whole or to a component system, it can address either improvement or new design. When the goal is improvement, analysis is carried out on the basis of the product system or component system(s). When the goal is design, analysis is based either on required specifications and plans or on design specifications.

PRODUCTION SYSTEMS

Functions

Production systems exist to make products. A production system is made up of people, equipment, and the energy to run the equipment. It includes functions that people should carry out and functions that machines should carry out. The mutual interaction among such functions transforms raw materials into finished goods.

Ease of Application

A production system's function of transforming materials facilitates function analysis because functions are reflected via changes in the materials. In addition, system behavior is stable and function analysis is easy because a production system is a conglomerate of static, dynamic, and control systems.

Modes of Application

When function analysis is applied to a production system, it may be applied to (a) the individual processes or series of processes that make up the system, (b) equipment or mechanical devices, or (c) human activities. When function analysis is applied to a production system or equipment system, moreover, it can address either improvement or design. Function analysis applied to systems of tasks or operations addresses itself to the question of improvement. (See Figure 5-3.)

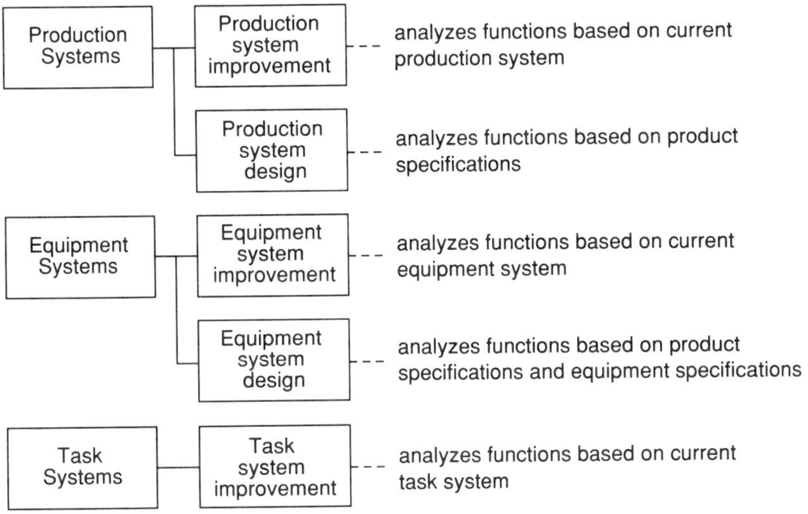

Figure 5-3. Modes of Application for Production Systems

MANAGEMENT SYSTEMS

Functions

Management systems exist to provide products or services to the market while ensuring profits.

As depicted in most organizational charts, the various organizational units within a company are the basic constituent components of its managerial system. Other components include assorted managerial tasks and office procedures within each organizational unit, as well as the people, information, and information-processing devices in the organization.

It follows that the functions a management system as a whole has to carry out are satisfied through the interaction of a variety of other functions, including organizational functions, managerial task functions, office procedure functions, functions needing to be carried out by humans, and functions needing to be carried out by machines.

Management system functions are information-gathering and process (including analysis) functions, as well as decision-making functions, and nearly all of them are carried out by people or by organizations.

Ease of Application

Because management system functions involve information and decision-making, the means for carrying them out are not visible, as they are for product parts or components. The intangibility of management systems makes application of function analysis more difficult than for product systems and production systems because different people have different perspectives on what functions need to be performed. At the same time, management systems are conglomerates of dynamic systems, control systems, and response; thus, they cannot be analyzed as systems with specific functions because their behavior changes with changing circumstances.

Modes of Application

Function analysis of management systems may be a specific system of support activities.

Whether applied to organizational or management systems, function analysis can be used either to improve an existing system or to design a new system.

Function analysis for improvement is based on current organizational systems, administrative systems, or task systems. In the case of design, function analysis is based on the goals and objectives of organizational systems, administrative systems, or task systems. (See Figure 5-4.)

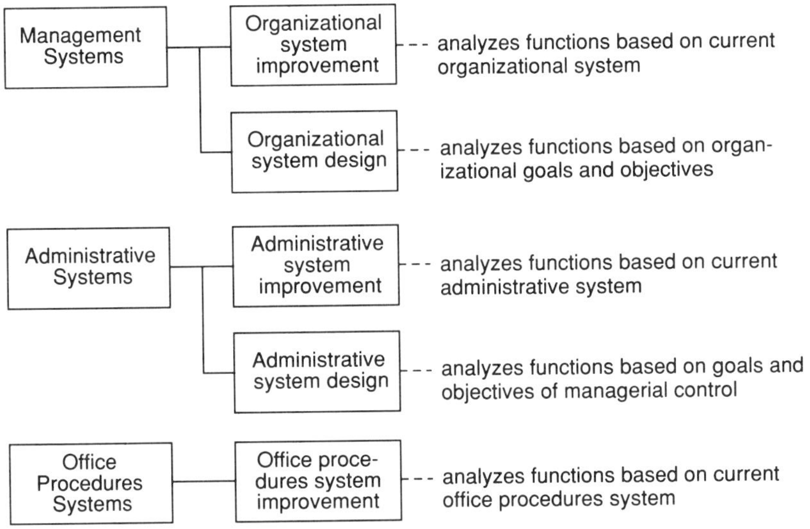

Figure 5-4. Modes of Application for Management Systems

SERVICE SYSTEMS

Functions

Service systems are systems that ensure profit while providing functions called "services" to the marketplace.

Services themselves are intangibles whose constituent components are the people who provide the services, the information for and about the services, and the machines and equipment related to the services. This means that people, information, and equipment all have functions to perform and that those individual functions interact to carry out the functions of the service system as a whole.

Even more than management system functions, the functions of a service system depend on people and in many cases consist of the responses of service providers to those being served. Not only are services intangible, their functions are equally intangible "purposive concepts." Services, in other words, are simply functions.

Ease of Application

The functions of a service system can be treated as the functions of the services themselves, and the human behavior and information that carry out those functions are invisible. Thus, like management system functions, service system functions are more difficult to analyze than product systems and production systems because different people have different perspectives on what functions need to be performed.

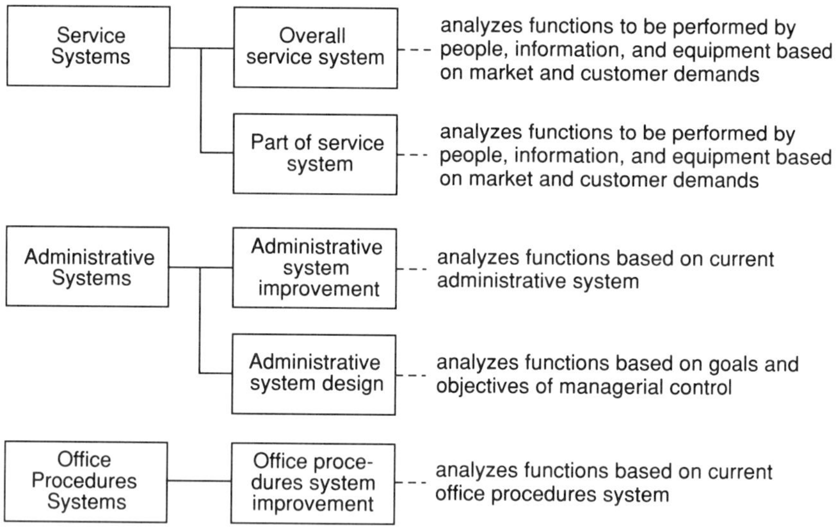

Figure 5-5. Modes of Application for Service Systems

Modes of Application

Function analysis may be applied to certain service systems as entities (such as retail service systems, restaurant service systems, or transport service systems). It is also sometimes applied to subsystems within service systems (for example, retail store sales systems or information provision systems). In addition, function analysis is sometimes applied to specific administrative support systems within service systems. (See Figure 5-5.)

6
Applying Function Analysis

PRODUCT IMPROVEMENT

The procedure for product improvement is outlined in Figure 6-1. A closer look at each of its steps follows.

Step 1: Select a Theme

Choose a product improvement theme. Product improvement means redesigning a currently produced product in order to obtain some advantage in terms of quality or cost. Product improvement therefore includes quality improvement involving design.

The first step in applying function analysis to product improvement is selecting a "theme," that is, a product that needs to be improved. Products encounter a variety of problems, of which quality and cost are common examples. A product chosen as the object of function analysis should suffer from a number of such problems.

The selection of a theme for product improvement should be carried out realistically, in accord with company circumstances. In this selection process, take into account the points shown on the checklist below.

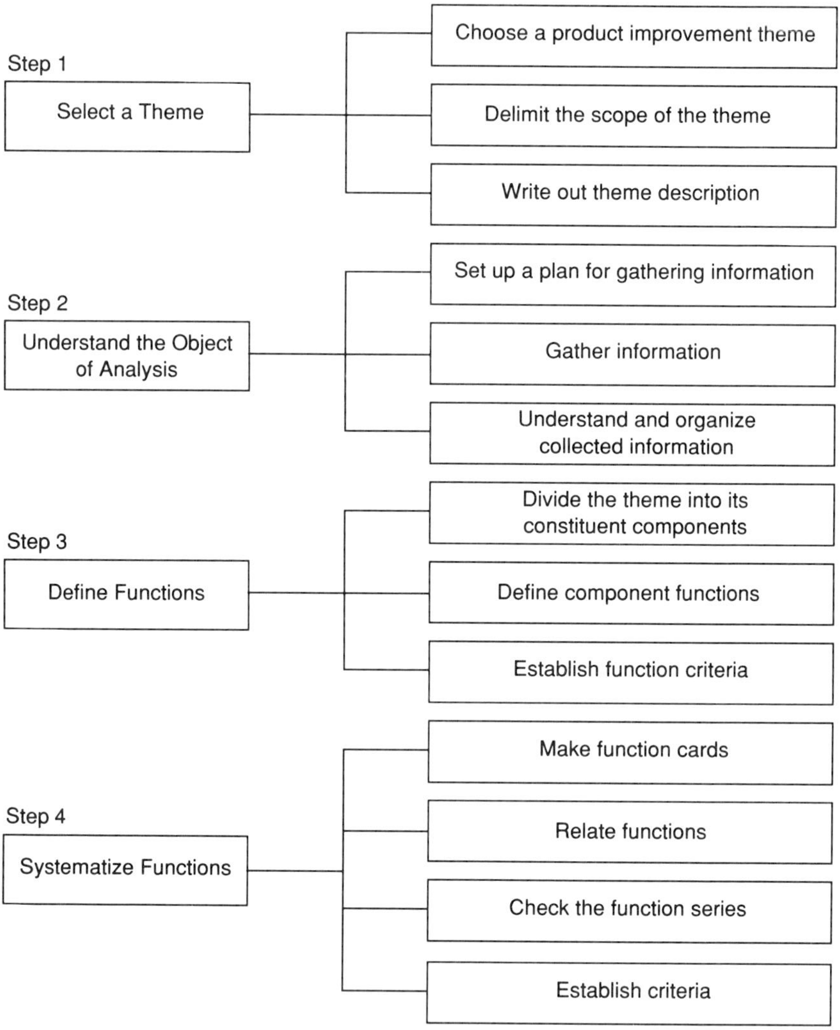

Figure 6-1. Product Improvement and Function Analysis Procedure

- Are sales of the product stable?
- Will sales increase?
- Is the profit margin high?

- Is the rate of return high?
- Are there constraints on product design changes?
- Has it been a long time since the product was designed?
- Has the product become difficult to manufacture?
- Is the cost ratio high compared to other products?
- Is there considerable room for cost reduction?
- Are there many marketplace quality problems due to the design?
- Are there many manufacturing quality problems due to the design?

Delimit the scope of the theme. The theme chosen for product improvement may be one of a number of types. Product improvement rarely means improving the body of the product itself (the body of a refrigerator, for example). More often, it involves improving one part of the product, that is, one of the product's subsystems. The reason for this is that product-related problems such as quality and cost reside not in the product as a whole but in specific parts of the product. When this is the case, only the relevant product part need be subject to improvement. Limiting the theme to specific parts makes it possible to carry out effective product improvement.

Once you have chosen a product improvement theme, you need to narrow its scope to clearly identify the domain in which function analysis will be used to make improvements.

Write out a theme description. When you have chosen a product improvement theme and limited its scope, write out a one-page summary description of the theme (see Figure 6-2).

Step 2: Understand the Object of Analysis

Set up a plan for collecting information. Once you have settled on a product improvement theme, the next step is to understand the object of analysis by collecting information relevant to the theme. "Information" means knowledge needed for

Function Analysis Worksheet 1

Description of Product Improvement Theme

1. Summary and scope of product improvement theme
   - Summary of product improvement theme (What is the theme?)

   - Scope of product improvement theme (What does the theme cover?)

2. Reasons for selecting product improvement theme (Why was this theme chosen?)
   -
   -
   -
   -
   -
   -

3. Product improvement theme objectives
   - Quality improvement objectives (How much better do we want to make quality?)

   - Cost improvement objectives (How far do we want to lower costs?)

   - Other improvement objectives (How much do we want to improve other aspects?)

- "Summary and scope of product improvement theme" — Write down what the theme is and what it encompasses. Use language that is simple, specific, and easy to understand.
- "Reasons for selecting product improvement theme" — Briefly list each reason for choosing the theme.
- "Product improvement theme objectives" — For quality-related objectives, write in numerical targets for quality characteristics, defect rates, customer complaints, etc. For cost-related objectives, write in cost reduction amounts, rate of cost reduction, etc.

Figure 6-2. Summary Description of Product Improvement Theme

accomplishing various goals. In the case of product improvement, you gather knowledge to improve the current product in terms of quality or cost.

Information is knowledge and therefore not something you can pick up and hold in your hand. This makes it crucial to establish a plan and assemble information that permits accurate understanding of the product improvement theme. Setting up a plan for information collecting means, first of all, identifying what information is needed. Then you select information sources, a timetable, collection methods, and individuals with specific responsibilities.

Table 6-1 lists the kinds of information that should be collected with respect to a product improvement theme. Information is assembled using forms for information to be collected (see Figure 6-3) and an information collection plan (see Figure 6-4).

Collect information. Information-collecting activities, carried out on the basis of the information collection plan, generally take time and effort, so adherence to the plan is essential. Keep the following points in mind when collecting information:

- Information is knowledge needed to attain a goal. Keep the goal in mind when collecting information.
- Information is sometimes tainted by conjecture or prejudice. Collect the facts.
- Information is intangible. Record clearly each piece of information you collect.
- Criteria may be fictitious. Be sure to grasp only true criteria.
- A variety of problems will be encountered. Be sure to grasp problems accurately.

Understand and organize collected information. All information collected provides important material for function analysis, which means that information collected must be fully

Table 6-1. Information to Be Collected for a Product Improvement Theme

Usage information. What demands, constraints, or problems are there regarding product use?
- Users, purposes, user demands
- Technical, social use environments
- Usage problems, accidents, marketplace claims

Sales information. What demands, constraints, or problems are there regarding product sales?
- Sales points, product differentiation
- Distribution demands, constraints
- Sales-related problems, accidents, claims

Design information. What product specifications, constraints, or problems are there regarding design?
- Performance, reliability, operability, maintainability, safety, and other specifications
- Current product design (drawings), parts makeup, materials
- Design history (design time frame, design changes, etc.)
- Design conditions (constraints, rules, standards, patents, etc., applying to design changes)
- Design problems (technical issues, design errors, etc.)

Procurement information. What demands, constraints, or problems are there regarding procurement of product materials?
- Procurement demands, conditions
- Items procured, sources, methods
- Procurement problems

Manufacturing information. What is the current state of manufacturing? Any constraints? Problems?
- Production volume, production capacity, equipment capabilities, manufacturing processes, work methods, standard times, efficiency
- Manufacturing problems (concerning capacity, quality, operations)

Cost information. What is the cost situation?
- Parts costs, processing costs, supplier costs, usage costs

Function Analysis Worksheet 2

Information to Be Collected

1. Usage information (usage-related demands, constraints, or problems)
   -
   -
   -
2. Sales information (sales-related demands, constraints, or problems)
   -
   -
   -
3. Design information (design-related specifications, constraints, problems)
   -
   -
   -
4. Procurement information (materials- and procurement-related demands, constraints, problems)
   -
   -
   -
5. Manufacturing information (current state of manufacturing, constraints, problems)
   -
   -
   -
6. Cost information (current cost situation)
   -
   -

- "Usage information" — List information for understanding usage ideal, constraints, problems.
- "Sales information" — List information for understanding sales constraints, problems.
- "Design information" — List information for understanding design ideal, constraints, problems.
- "Procurement information" — List information for understanding current procurement situation, problems.
- "Manufacturing information" — List information for understanding current manufacturing situation, problems.
- "Cost information" — List information for understanding current cost situation.

Figure 6-3. Form for Collecting Information

Function Analysis Worksheet 3

Information Collection Plan					
Information Type	Information to Be Collected	Source(s)	Collection Schedule	Collection Method	Person in Charge
1. Usage information	• • •				
2. Sales information	• • •				
3. Design information	• • •				
4. Procurement information	• • •				
5. Manufacturing information	• • •				
6. Cost information	• • •				

- "Information to Be Collected" — Transfer information to be collected from Function Analysis Worksheet 2
- "Source(s)" — Determine the appropriate departments or persons from whom each type of information is to be collected and write them in as sources.
- "Collection Schedule" — Set deadlines for the collection of each type of information and write in the corresponding date.
- "Collection Method" — Decide how each type of information is to be collected and write in the collection method.
- "Person in Charge" — Decide who is going to collect each type of information and write in that person's name.

Figure 6-4. Information Collection Plan

understood. Information that is incomplete (or that can be misconstrued) must be collected again. Besides facilitating function analysis, a full understanding of the information gives you an understanding of the product improvement theme that constitutes the object of analysis.

Collected information should be filed for ready access and ease of use in subsequent analyses.

Step 3: Define Functions

Divide the theme into its constituent components. A product improvement theme will encompass either an entire product or a product subsystem. In either case, there will be a variety of functions to perform.

To make it easier to define the various functions that must be performed, divide the product improvement theme into its constituent components and then define the function of each component. The size of the units into which constituent components are broken down will vary according to the scope and complexity of the theme and the improvement goals. The degree of detail of defined functions, moreover, will depend on the fineness of division into constituent components.

Figure 6-5 gives the conceptual framework of division into constituent components.

Define the component functions. Using verb-noun combinations, define the functions to be performed by the components you have identified. Generally speaking, individual components perform functions not on their own but as specific parts of the overall product improvement theme. In this situation, we define the function to be played by each component in the context of the system as a whole. We also need to be aware, as we define functions, that a single component may have multiple functions to perform.

Improvement goals \ Theme complexity	Complex structure with many parts	Simple structure with few parts
Significant results expected	• Large subassembly units treated as constituent components	• Small subassembly units treated as constituent components
Significant results not expected	• Small subassembly units treated as constituent components	• Parts treated as constituent components

• In any case, the number of constituent components should range from 10 to 30.

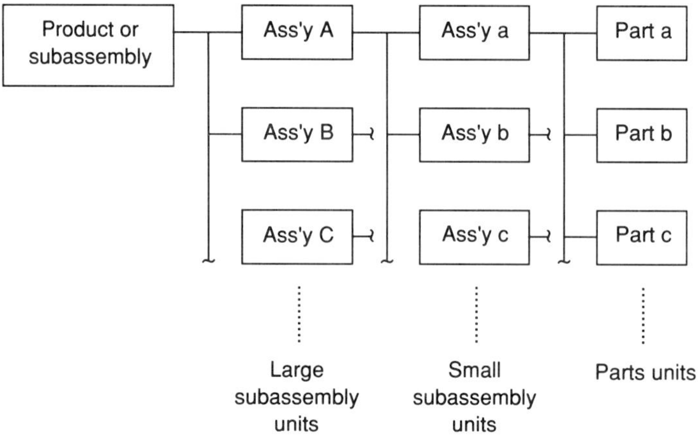

Figure 6-5. Division into Constituent Components

Below are important points to keep in mind when defining functions to be performed by constituent components:

- Define functions simply, with verb-noun expressions.
- Use easy language anyone can understand.
- Use specific terms not subject to misinterpretation.

- Define functions without referring to specific names of parts or constituent components.
- Avoid using proper nouns in definitions.
- Use transitive verbs (i.e., verbs that take direct objects).
- Use abstract function expressions whenever possible.
- Use easily quantifiable function expressions whenever possible.

Establish the function criteria. All defined functions are subject to various constraints on the ways and degrees to which they can be achieved, and the next step is to determine those constraining conditions — or criteria — and relate them to each function. In the case of product improvement, such criteria will ordinarily be determined by product specifications, but current specifications are not always suitable from the customer's point of view. This makes it important to establish criteria appropriate to the customer's perspective on the basis of information collected in step 2.

While these criteria refer to product specifications, they also express product quality and quality level, and therefore constitute a prerequisite for product improvement.

Strict criteria limit the scope of product improvement, whereas relatively lax criteria facilitate such improvement. On the other hand, narrow criteria make it easier to assess improvement, while liberal criteria are vague and thus make evaluation difficult.

Keep the previous issues in mind when you establish criteria for functions (See Figure 6-6).

Step 4: Systematize Functions

Make function cards. Transfer each function identified through function definition onto a card like the one shown in Figure 6-7. You should have the same number of cards as identified functions.

Function Analysis Worksheet 4

No.	Part or Component	Diagram	Function		
			Verb	Noun	Criteria

Figure 6-6. Establishing Criteria for Functions

On each card, in the space labeled "function," record the defined function in large block letters. In the space labeled "component," write either the name or number of the constituent component whose function is defined.

Relate the functions. Select a function card at random and look for the goal of that function by directing the question *why* to the function on the card. Identify the function matching the

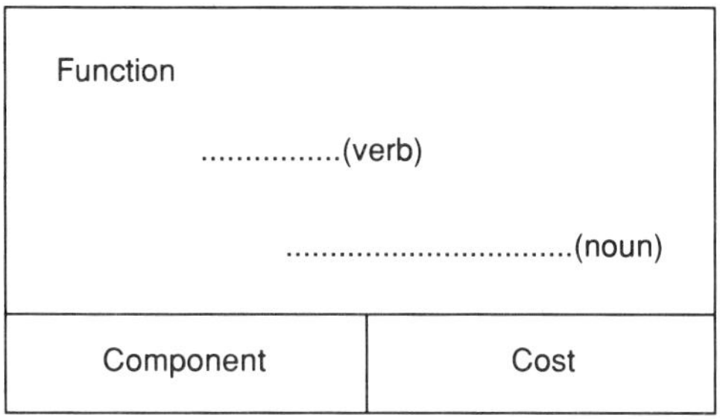

Figure 6-7. Function Card

response to that question and place its function card to the left of the first card (see Figure 6-8).

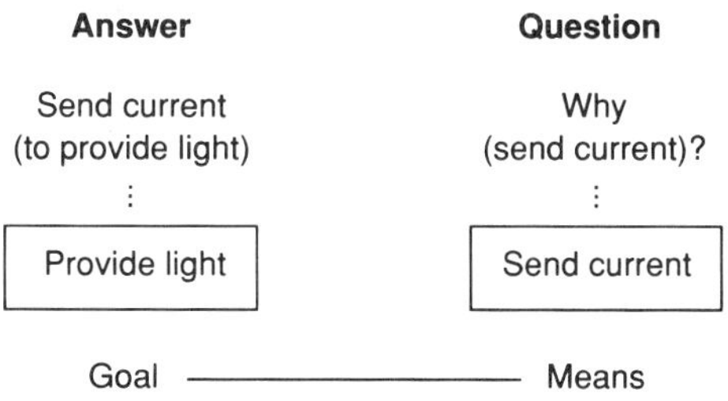

Figure 6-8. Relating Functions

If the "response" function is recorded on one of the cards you have already made, simply lay it down on the left. If it is not recorded, make up a new card and place the new card to the left of the "question" function card. Continue searching for goal functions by repeating this procedure until a "response" function falls outside the scope of the improvement theme. You systematize functions by carrying out the above procedure with all the function cards.

If the process just described fails to produce clear goal functions, revise your function definitions.

Check the function series. When all defined functions have been positioned in the function series, verify the appropriateness of the series. Apply the following questions in checking the function series:

- Can the goal functions be reliably satisfied by the means functions?
- Are the means functions unnecessary if the goal functions are unnecessary?

Means functions are inadequate — they "leak" — if they cannot reliably satisfy goal functions. A function needed as a means to another function is incorrectly positioned, moreover, if its ostensible goal function is unnecessary. In either case, you need to add functions, change the position of functions, or revise the function series.

In other instances, customer demands or product problems may result in inadequate functions. Here, too, functions need to be tracked down and the function series corrected.

Establish criteria. On the completed function family tree, write in criteria corresponding to all the functions. Copy criteria established in the function definition stage above the functions they constrain. Criteria for high-order functions, however, have not yet been established on the function family tree. Fill in these conditions based on usage and design information (see Figure 6-9).

PRODUCT DESIGN

The procedure for product design and function analysis is outlined in Figure 6-10. A closer look at each step follows.

Step 1: Understand the Object of Analysis

Collect information about the object of analysis. Product design means drawing up and determining the specific details

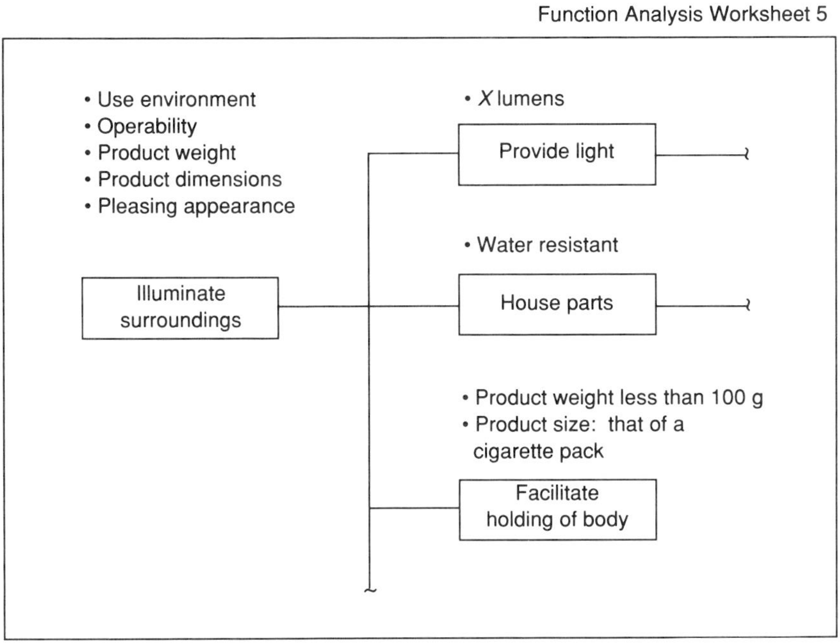

Figure 6-9. Establishing Criteria

of a new plan for a product that will satisfy the demands of function, quality, and cost.

Product design typically proceeds on the basis of customer or in-house demands incorporated in a planning document or development specifications sheet. It may be, however, that a written plan or development specifications sheet does not accurately include all requirements. Indeed, product design is often carried out on the basis of inadequate written plans or specifications sheets.

While it is, of course, important to check the contents of planning documents and development specifications sheets, applying function analysis to product design also requires that you collect necessary information and fully understand what is needed of the product to be designed.

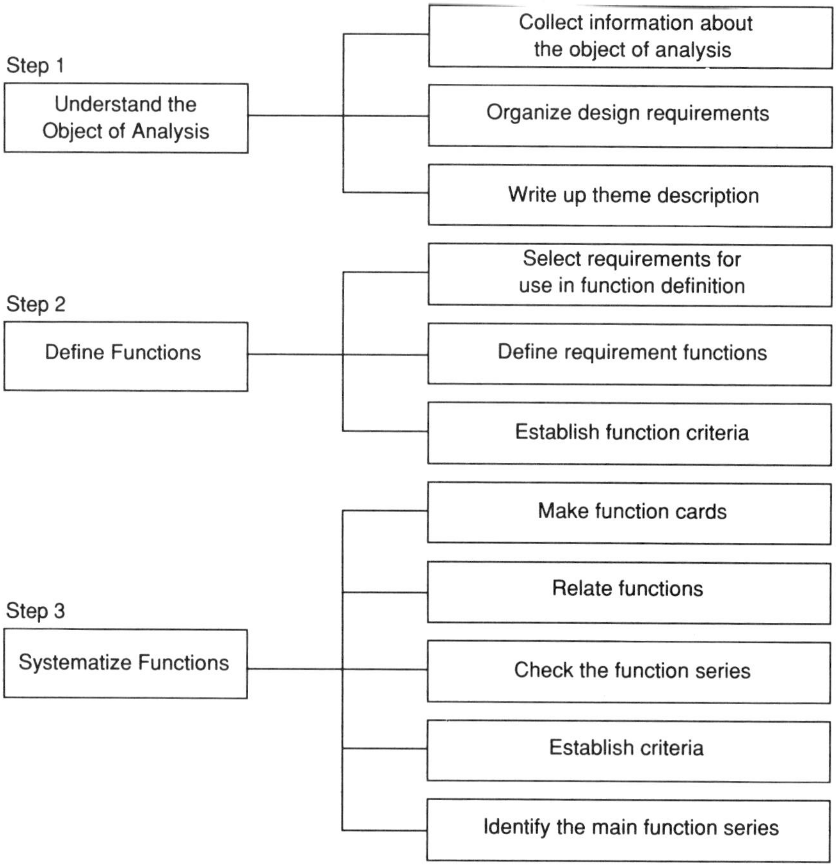

Figure 6-10. Product Design and Function Analysis Procedure

When applying function analysis to product design, collect information on requirements such as those shown in Table 6-2.

Organize the design requirements. After they are organized, the various requirements collected in the form of information can be integrated into the design. Not all requirements need be incorporated into the design, however. Decide on final design requirements after making trade-offs among factors such as function (including quality), production volume, and cost.

Table 6-2. Information to Be Collected for Product Design Requirements

Usage requirements
- Requirements concerning users (age, sex, social class, etc.) and purposes
- Requirements concerning use environments (temperature, humidity, location, etc.)
- Requirements concerning performance (satisfaction of functions)
- Requirements concerning reliability (maintenance of function)
- Requirements concerning maintainability (ease of function restoration, repair)
- Requirements concerning operability (ease of operation)

Sales requirements
- Requirements concerning sales points, product differentiation
- Requirements concerning distribution and transport (routes, methods, units)
- Items required by pricing (standard prices)
- Requirements concerning sales periods and volumes

Design requirements
- Requirements concerning legal regulations and various standardization issues
- Cost items (fabrication costs and allowable investment costs)
- Design deadline requirements

Manufacturing requirements
- Requirements concerning production equipment
- Requirements concerning place of (and space for) manufacturing
- Requirements concerning required staffing and fabrication labor costs

Procurement requirements
- Requirements concerning in-house vs. external fabrication

These requirements include all demands and constraints placed on product design (see Figure 6-11).

Once all design requirements have been determined, draw up a requirements list like the one shown in Figure 6-12.

Write out a theme description. When you have settled on design requirements, write out a product design theme description based on those requirements. (See Figure 6-13.) A prerequisite to product design, such a description will incorporate

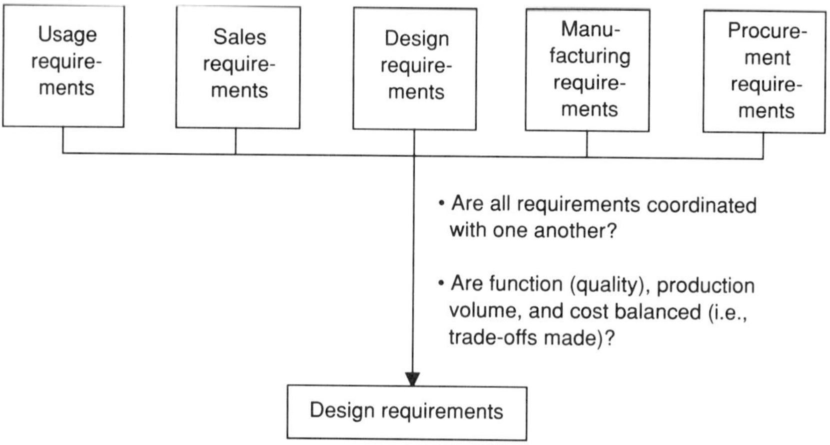

Figure 6-11. Establishing Design Requirements

Function Analysis Worksheet 1

Requirements List

Usage requirements
- Users ....
- Use environments ....
- Performance ....
- Reliability ....
- Maintainability ....
- Safety ....

Sales requirements
- Differentiation and sales points ....
- Distribution and transport ....
- Pricing (standard price) ....
- Sales periods and volumes ....

Figure 6-12. Product Design Requirements List

information corresponding to a written plan or development specifications sheet.

Function Analysis Worksheet 2

Product Design Theme Description

1. Product name
2. Technical demands and parameters (based on requirements, but more specific)
   - Product users and user goals
   - Place and environment of product use
   - Product performance and reliability
   - Product operability and maintainability
   - Product safety
   - Product sales criteria
   - Product design criteria
   - Product fabrication and procurement criteria
3. Economic demands and parameters (based on requirements)
   - Product sales price (standard price and estimated actual price)
   - Product cost (fabrication cost and sales costs)
   - Development investment costs
4. Other product design-related demands and parameters (based on requirements)
   - Production and sales volumes (number of units planned)
   - Design deadlines (draft design schedule, etc.)
   - Other

Figure 6-13. Product Design Theme Description

Step 2: Define Functions

Select the requirements for use in function definition. Functions are concepts — product actions purposefully assigned to products awaiting design. Clearly identifying what those functions are is the starting point of the design process. From among design requirements already identified, therefore, you select those requirements representing functions the product needs to perform (Table 6-3).

Table 6-3. Selection of Requirements to Be Treated as Product Functions

Category / Requirements	Functional Require-ments	Sales Require-ments	Economic Require-ments	Manufac-turing Require-ments	Deadline Require-ments
Purposes	★				
Use environment	★				
Performance	★				
Reliability	★				
Maintainability	★				
Operability	★				
Safety	★				
Differentiation	★				
Distribution and transport		★			
Price			★		
Sales period					★
Sales volume		★			
Regulations	★				
Standardization	★				
Cost			★		
Design period					★
Equipment used				★	
Place of fabrication				★	
Needed personnel				★	
In-house vs. outside fabrication				★	

The requirements you have identified will encompass a variety of demands made on the product. These demands can be classified roughly as requirements relating to product function, product sales, economic issues, manufacturing, and delivery. (See Table 6-4.)

Table 6-4. Criteria for Classifying Requirements

Category	Principles
Functional requirements	• Requirements relating directly to physical conditions or to functions demanded by technical issues. (These requirements are a principal aim of product design and are prerequisite to function definition.)
Sales requirements	• Design requirements other than technical issues: requirements involving product sales, distribution, or transport. (These requirements may be constrained by product design.)
Economic requirements	• Design requirements relating to product price, costs, and other economic issues. (These requirements are taken up in the design process as constraints on — or trade-offs with — functional requirements.)
Manufacturing requirements	• Requirements relating to product manufacture: design requirements that constrain the physical makeup of the product. (These requirements are addressed in the product design stage when the product's physical makeup is determined.)
Delivery requirements	• Design requirements relating to deadlines for product design, manufacturing, and sales. (These requirements are addressed in terms of product design deadlines and other time restrictions.)

Of these various demands, product function-related requirements are those that should be treated as product functions. The other demands need to be dealt with in the design process as factors that, for the most part, constrain product design without impinging on product function. Use the function analysis worksheet shown in Figure 6-14 to summarize requirements selected for the purpose of function definition.

Function Analysis Worksheet 3

Requirement	Description
Purpose	
Use environment	
Performance	
Reliability	
Maintainability	
Operability	

(etc.)

Figure 6-14. Requirements Needed for Function Definition

Define the requirement functions. For each requirement selected for function definition, define the function to be performed by the product with a verb-noun pair.

A requirement is expressed in narrative terms, but that narrative expression must accurately convey the function to be performed by the product. (See Figure 6-15.)

Function Analysis Worksheet 4

Requirements	Functions		
	Verb	Noun	Criteria
(Purposes) This pocket flashlight is to be used by women and children to illuminate walkways, etc., when they return home after dark. It fits in the pocket or purse and is both attractive and easy to use.	illuminate facilitate retain	surroundings storage attractiveness	*X* lux for *Y* meters around cigarette pack-sized holds its appeal to women and children
(Use environment) Used at night, out of doors, and even on rainy days; the body can be attached permanently to provide a stationary light source.	facilitates carrying of prevent hold	body entry of water body	can be held in one hand water-resistant
(Performance) Light operates at *x* watts; flash uses 2 size A batteries.	provide houses	light batteries	*X* watts

(etc.)

Figure 6-15. Defining Requirement Functions

Tips for Defining Functions

- Gain an adequate understanding of each requirement
- Treat each requirement as a function the product needs to perform
- Understand each requirement in terms of its essence, not in terms of the way it is expressed

Establish the function criteria. All defined functions are subject to various constraints on the ways and degrees to which they can be achieved, and the next step is to determine those constraining conditions — or criteria — and relate them to each function. In terms of the various requirements we have already classified, criteria are critical factors that may, for example, (a) limit the scope within which product functions are performed, (b) limit the means by which functions can be performed, or (c) constrain the physical parameters of the product. We therefore base such criteria on function, sales, and manufacturing requirements.

In function analysis, we establish, or specify, such criteria — ordinarily referred to in product design as development specifications or design specifications — in relation to functions expressed by verb-noun expressions.

A key question when relating individual criteria to functions is the following: Which criteria constrain the achievement of which functions? (See Figure 6-16.)

Criterion	**Corresponding Function**
X lux for *Y* meters	Lux is a function of luminous intensity and therefore constrains — and corresponds to — the "illuminate surroundings" function.
Attractive to women and children	Attractiveness is a function of visual appearance derived from design, color, etc., and therefore corresponds to the "holds appeal" functions.
Water resistant	Water resistance is a function of product waterproofness and therefore corresponds to the "prevents entry of water" function.

Figure 6-16. Relating Criteria to Functions

Criteria are function-, sales-, or manufacturing-related demands that indicate limits on the performance of product function or on physical product parameters. This means we need to recheck requirements or written plans whenever criteria are unclear.

Step 3: Systematize Functions

Make function cards. Transfer each function identified through function definition onto a card like the one shown in Figure 6-17, so that you have the same number of cards as identified functions.

In the space labeled "function" on each card, record the defined function in large block letters. For product design, leave the spaces labeled "component" and "cost" blank.

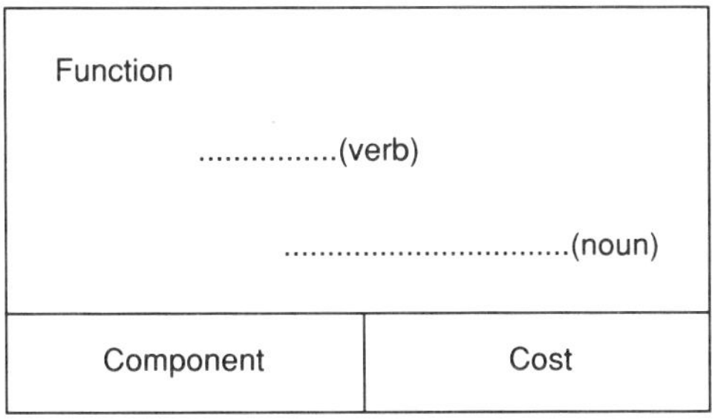

Figure 6-17. Function Card

Relate the functions. Select a function card at random and look for the goal of that function by directing the question *why* to the function on the card. Identify the function matching the response to that question and place its function card to the left of the first card (see Figure 6-18).

If the "response" function is recorded on one of the cards you have already made, simply lay it down on the left. If it is not recorded, make up a new card and place the new card to the

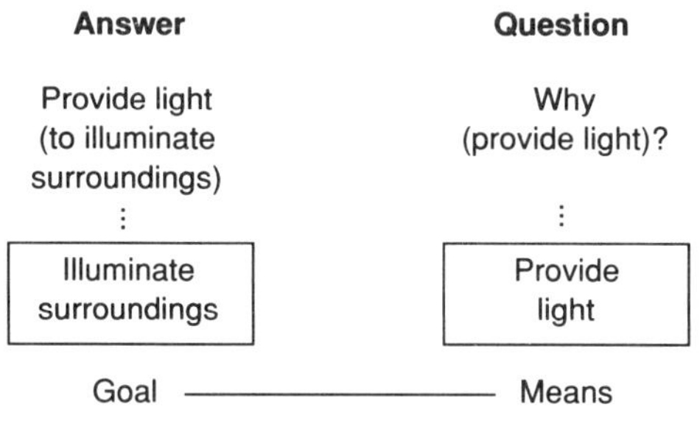

Figure 6-18. Relating Functions

left of the "question" function card. Continue searching for goal functions by repeating this procedure until a "response" function falls outside the scope of the design target. Systematize the functions by carrying out the above procedure with all the function cards.

If the process just described fails to produce clear goal functions, revise your function definitions.

Check the function series. When all defined functions have been positioned in the function series, verify the appropriateness of the series. Apply the following questions in checking the function series:

- Can the goal functions be reliably satisfied by the means functions?
- Are the means functions unnecessary if the goal functions are unnecessary?
- Are basic functions the product must perform incorporated without "leaks"?
- Can the product design concept be easily understood from the function series?

Functions should be added or corrected — or the function series revised — if either (1) means functions cannot reliably satisfy goal functions or (2) means functions serve any but necessary goal functions.

Functions or the function series should be revised, too, if the basic functions expected of the product are not adequately accounted for or if product design concepts are not easily understandable.

Establish the criteria. On the completed function family tree, write in criteria corresponding to all the functions. If all criteria established in the function definition stage were placed on the function family tree, check to see that all criteria originating in design requirements figure on the chart as well. (See Figure 6-19.)

Function Analysis Worksheet 5

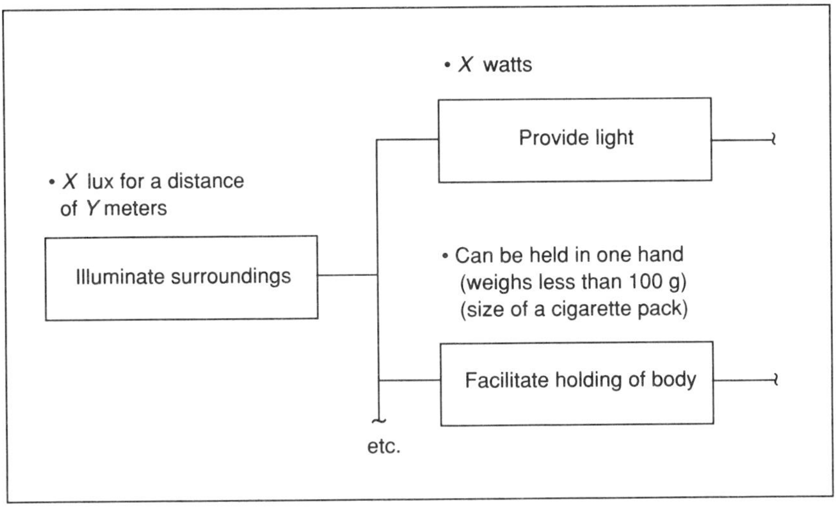

Figure 6-19. Establishing Criteria

Identify the main function series. A completed function family tree provides a technical basis for proceeding with product design. Functions that still need to be designed into the product are given high visibility by the completion of a function family tree.

This means that a function family tree — a result of function analysis — becomes a basis both for creating the means to satisfy further functions and for completing design drafts. At the same time, functions shown on the function family tree are of various types. The overall product design is completed by identifying the main function series, designing the means to accomplish those functions, and then designing the means to accomplish functions in the supporting function series.

Thus, the next step is to identify the main function series on the function family tree. By "main function series" we mean the function connections on a function family tree that prescribe the basic method of designing the product. Basically, there is only one main function series, but in some cases there may be two. (See Figure 6-20.)

PROCESS IMPROVEMENT

The procedure for process improvement through function analysis is outlined in Figure 6-21. A closer look at each step follows.

Step 1: Select a Theme

Choose a process improvement theme. Process improvement means modifying the production system to obtain better performance in quality, cost, or delivery.

The first step in applying function analysis to process improvement is theme selection: What process series needs to be improved?

Processes involve a variety of problems, including quality, cost, and the ability to meet deliveries, and the process selected as the target of function analysis should suffer from such problems. The selection of a theme for process improvement should be carried out realistically, in accord with company circumstances. The points shown on the following checklist should be taken into account in this selection process:

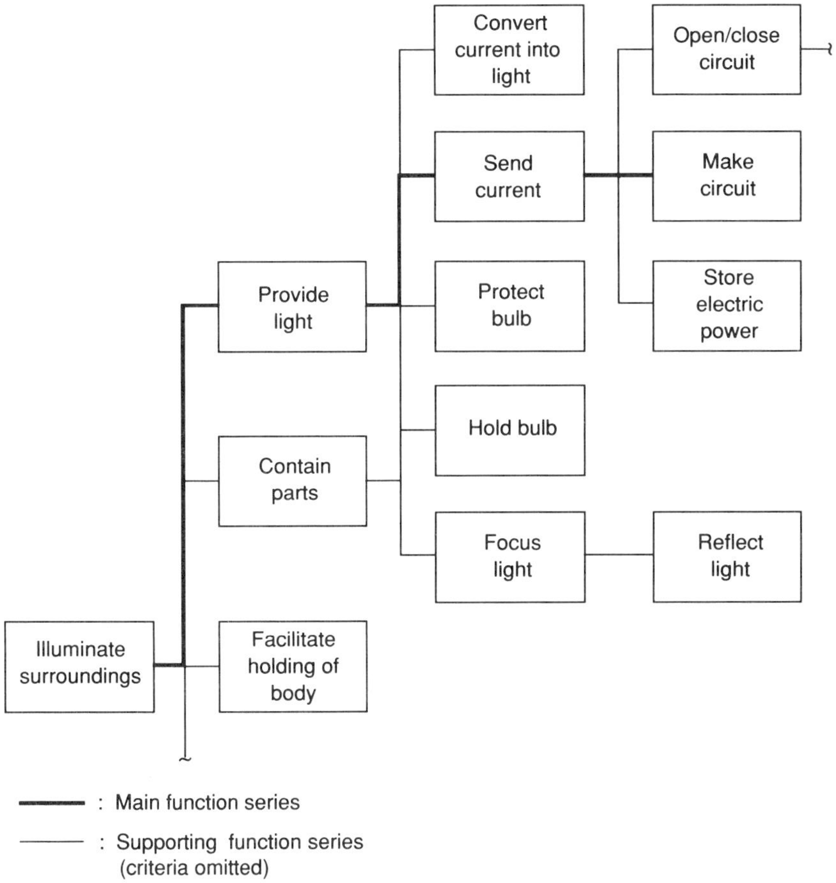

Figure 6-20. Identifying Main Function Series

- Does the process suffer from many problems?
- Does process cost too much relative to function?
- Is there significant room for cutting back processing costs?
- Is the process constantly responsible for missed delivery deadlines?
- Will production volume for the process remain stable?

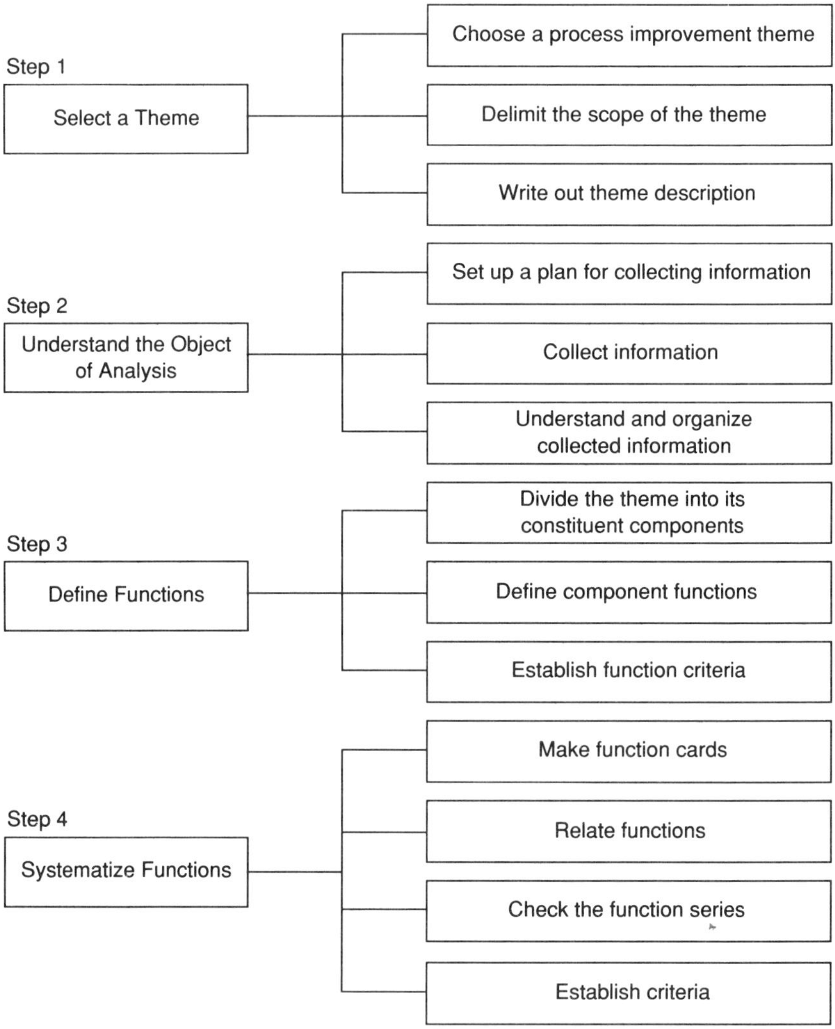

Figure 6-21. Process Improvement and Function Analysis Procedure

- Will production volume for the process rise?
- Does the process have common elements that will multiply the effects of improvement?

- Are there numerous constraining criteria involved in changing the process?
- Have the same processing methods been used for years?
- Are there operability or safety problems in the process?

Delimit the scope of the theme. The theme chosen for product improvement may be one of a number of different types. Process improvement rarely involves improving all process series — more often, it means targeting specific ones. The reason for this is that process problems related to quality, cost, and delivery reside in specific processes. When this is the case, only those specific process series need be subject to improvement. Limiting the theme to specific processes makes it possible to carry out effective process improvement.

Once you have chosen a process improvement theme, you need to narrow its scope to identify clearly the domain in which function analysis will be used to make improvements.

Write out a theme description. When you have chosen a process improvement theme and limited its scope, write out a one-page summary description of the theme. (See Figure 6-22.)

Step 2: Understand the Object of Analysis

Set up a plan for collecting information. Once you have settled on a process improvement theme, the next step is to understand the object of analysis by collecting information relevant to the theme. "Information" means knowledge needed for accomplishing various goals. In the case of process improvement, you gather knowledge to improve the current process in terms of quality, cost, delivery, and so on.

Information is knowledge and therefore not something you can pick up and hold in your hand. This makes it crucial to establish a plan and assemble information that permits accurate understanding of the process improvement theme. Setting up a plan for information collecting means, first of all, identifying

Function Analysis Worksheet 1

Description of Process Improvement Theme

1. Summary and scope of process improvement theme
   - Summary of process improvement theme (What process is to be improved?)

   - Scope of process improvement theme (What does the theme process cover?)

2. Reasons for selecting process improvement theme (Why was this process chosen as a theme?)
   -
   -
   -
   -
   -
   -

3. Process improvement theme objectives
   - Quality improvement objectives (How much better do we want to make quality?)

   - Cost improvement objectives (How far do we want to lower costs?)

   - Delivery improvement objectives (How much do we want to shorten lead time?)

- "Summary and scope of process improvement theme" — Write down what the theme is and what it encompasses. Use language that is simple, specific, and easy to understand.
- "Reasons for selecting process improvement theme" — Briefly list each reason for choosing the theme.
- "Process improvement theme objectives" — For quality-related objectives, use numerical targets for quality characteristics, defect rates, customer complaints, etc. For cost-related objectives, write in cost reduction amounts and rate of cost reduction. For delivery-related objectives, fill in production capacity and production lead time, etc.

Figure 6-22. Summary Description of Process Improvement Theme

what information is needed, and then selecting information sources, a timetable, collection methods, and individuals with specific responsibilities.

Table 6-5 shows the kind of information that should be collected for a process improvement theme. Information is assembled using forms for information to be collected (see Figure 6-23) and an information collection plan (see Figure 6- 24).

Information-collecting activities, carried out on the basis of the information collection plan, generally take time and effort, so adherence to the plan is essential. Keep the following points in mind when gathering information:

- Information is knowledge needed to attain a goal. Keep the goal in mind when collecting information.
- Information is sometimes tainted by conjecture or prejudice. Collect the facts.
- Information is intangible. Record clearly each piece of information you collect.
- Criteria may be fictitious. Be sure to grasp only true criteria.
- A variety of problems will be encountered. Be sure to grasp problems accurately.

Understand and organize the collected information. All information collected provides important material for function analysis and this makes it essential that information collected be fully understood. Inadequately understood information must be collected again.

A full understanding of the information also means an understanding of the product improvement theme that constitutes the object of analysis.

Collected information should be organized for ready access and ease of use in subsequent analyses.

Table 6-5. Information to Be Collected for a Process Improvement Theme

Product information (What specifications and quality are demanded of the product?)
- Product specifications
- Product drawings
- Parts drawings

Process information (What is the current state of the process?)
- Process analysis chart
- Process layout diagram
- Tools, dies, and equipment used
- Production volume, production capacity

Operational information (What is the current state of operations?)
- Work analysis charts, standard operations methods
- Operation layout chart

Productivity information (What is the current state of production?)
- Standard times
- Equipment operating rates, work efficiency
- Current labor requirements

Cost information (Give breakdown of processing costs)
- Reject rate, materials yield
- Processing costs by process

Other information
- Information on packaging, storage, and transport
- Problems involving manufacturing processes, operations

Step 3: Define Functions

Divide the theme into its constituent components. A process improvement theme will ordinarily correspond to a specific series of processes. Each of the processes in a process series will have various functions it needs to perform. To make it easier to define the various functions that must be performed, divide the process improvement theme into its constituent components and then define the function of each component.

Function Analysis Worksheet 2

Information to Be Collected

1. Product information (What specifications and quality are demanded of the product?)
   -
   -
   -
2. Process information (What is the current state of the process?)
   -
   -
   -
3. Operational information (What is the current state of operations?)
   -
   -
   -
4. Productivity information (What is the current state of production?)
   -
   -
   -
5. Cost information (Give breakdown of processing costs)
   -
   -
6. Other information
   -
   -

- "Product information" — List information for understanding product specifications and quality issues that are technical prerequisites of the fabrication process.
- "Process information" — List information for understanding the current state of the process.
- "Operational information" — List information for understanding the current state of operations.
- "Productivity information" — List information for understanding the current state of productivity.
- "Cost information" — List information for understanding the state of processing costs.
- "Other information" — List other process-related information.

Figure 6-23. Form for Collecting Information

Function Analysis Worksheet 3

Information Collection Plan					
Information Type	**Information to Be Collected**	**Source(s)**	**Collection Schedule**	**Collection Method**	**Person in Charge**
1. Product information	• • •				
2. Process information	• • •				
3. Operational information	• • •				
4. Productivity information	• • •				
5. Cost information	• • •				
6. Other information	• • •				

- "Information to Be Collected" — Transfer information to be collected from Function Analysis Worksheet 2.
- "Source(s)" — Determine the appropriate departments or persons from whom each type of information is to be collected and write them in as sources.
- "Collection Schedule" — Set deadlines for the collection of each type of information and write in the corresponding date.
- "Collection Method" — Decide how each type of information is to be collected and write in the collection method.
- "Person in Charge" — Decide who is going to collect each type of information and write in that person's name.

Figure 6-24. Information Collection Plan

There are various levels of units into which process series may be broken down. The particular level at which we define the functions of constituent components will depend on the goals of process series improvement. Figure 6-25 and Table 6-6 show how units are affected by improvement goals.

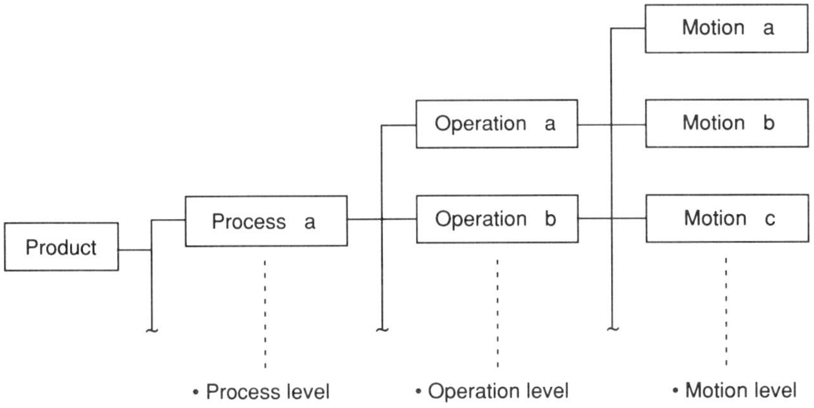

Figure 6-25. Process Series Breakdown Structure

Table 6-6. Process Series Improvement Goals and Breakdown Structure

Improvement goals	Units	Approach to analytic units
Process series improvements	Process	Function definition focuses on physical and chemical change processes leading from raw materials to finished products.
Work methods improvements	Operation level	Function definition focuses on the work of operators and machines making up the process.
Motion methods	Motion level	Function definition focuses on the series of motions or movements that make up the operation.

Define the component functions. Using verb-noun combinations, define the functions to be performed by the components you have identified. In the case of a process improvement theme, each function to be performed by a constituent component will serve to change some target or "object." Defining the function of each constituent component, then, becomes a matter of (a) checking the state of that object in the process or operation before and after application of the function and then (b) defining the change in terms of a function.[1] Figure 6-26 illustrates these circumstances in terms of a feed-packing process.

Tips for Defining Functions

- Use simple, verb-noun definitions. If necessary, add words to the noun component.
- Define functions using simple terms not subject to misinterpretation.
- Define functions without referring to specific names of processes or operations.
- Use transitive verbs (i.e., verbs that take direct objects).
- Verify the significance of the change and define it accurately in terms of a function.

Establish the function criteria. All defined functions are subject to various constraints on the ways and degrees to which they can be achieved, and the next step is to determine those constraining conditions — or criteria — and relate them to each function. In the case of process improvement, these criteria ordinarily correspond to levels of product quality that need to be built into the process. When product quality has already been clearly determined, criteria can be set merely by relating the quality levels to functions. When product quality is not

Scattered parts (before)

Assembly process — "assemble parts" (function)

Assembled parts (after)

Grease-covered parts (before)

Degreasing process — "remove grease" (function)

Degreased parts (after)

(Powdered) feed and sacks (before)

Packing process — "put in feed" (function)

Feed in sacks (after)

Figure 6-26. Functions as Changes of State in a Feed-packing Process

clear, on the other hand, criteria must be set on the basis of product specifications.

Since current product quality is not necessarily appropriate, it is important to use information collected in step 2 and establish criteria by setting a suitable level of quality in terms of product specifications. (See Figure 6-27.)

These product quality criteria become essential prerequisites when one sets up other function-serving methods of processing or fabrication and then continues to make process or operations improvements. Strict criteria limit the scope of process or operational improvement, whereas lax criteria facilitate such improvement. On the other hand, while narrow criteria make it easier to assess improvement, liberal criteria are vague and thus make evaluation difficult.

Function Analysis Worksheet 4

Process and *What* Chart	Function	Criteria

Figure 6-27. Establishing Criteria Using Product Specifications

Keep the above issues in mind when you set product quality criteria for functions.

Step 4: Systematize Functions

Make function cards. Transfer each function identified by function definition onto a card like the one shown in Figure 6-28, so that you have the same number of cards as identified functions.

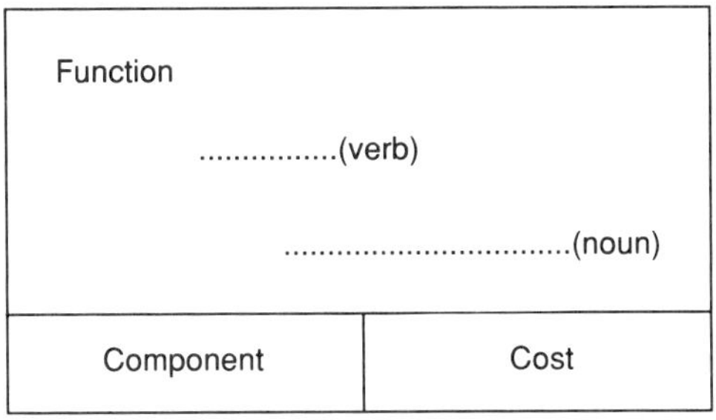

Figure 6-28. Function Card

In the space labeled "function" on each card, record the defined function in large block letters. In the space labeled "component," write either the name or number of the constituent component whose function is defined.

Relate the functions. Select a function card at random and look for the goal of that function by directing the question *why* to the function on the card. Identify the function matching the response to that question and place its function card to the left of the first card. (See Figure 6-29.)

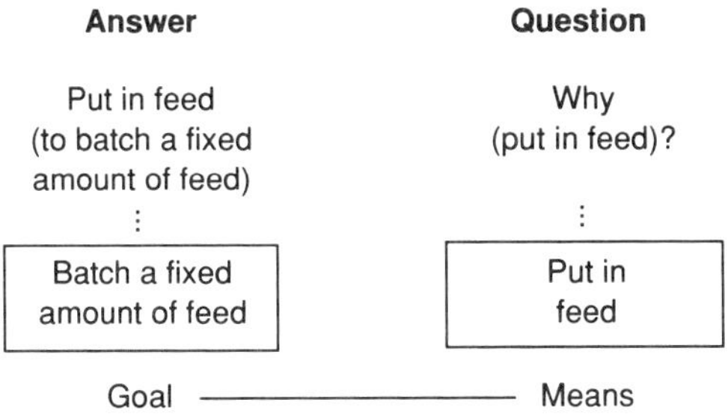

Figure 6-29. Relating Functions

If the goal function is not among already existing cards, make up a new card for it and place it to the left of the means function card. Continue searching for goal functions by repeating this procedure until a "response" function falls outside the scope of the improvement theme. You systematize functions by carrying out the above procedure with all the function cards.

There is a tendency, in analyzing process improvement themes, to think of the goal function as the next function. Although the goal function may indeed be the next function, this is not always the case. It is crucial to think of goal functions without regard for the current process sequence.

Check the function series. When all defined functions have been positioned in the function series, verify the appropriateness of the series. Apply the following questions in checking the function series:

- Can the goal functions be reliably satisfied by the means functions?
- Are the means functions unnecessary if the goal functions are unnecessary?

Means functions are inadequate — they "leak" — if they cannot reliably satisfy goal functions. A function needed as a means to another function is incorrectly positioned, moreover, if its ostensible goal function is unnecessary. Either case calls for adding functions, changing the position of functions, or revising the function series.

Establish the criteria. Write in the corresponding criteria above the functions on the completed function family tree. Copy criteria set in the function definition stage above the functions they constrain. Establish and record new high-order goal function criteria where needed. (See Figure 6-30.)

PROCESS DESIGN

The procedure for process design and function analysis is outlined in Figure 6-31. Each of its steps follows in detail.

Function Analysis Worksheet 5

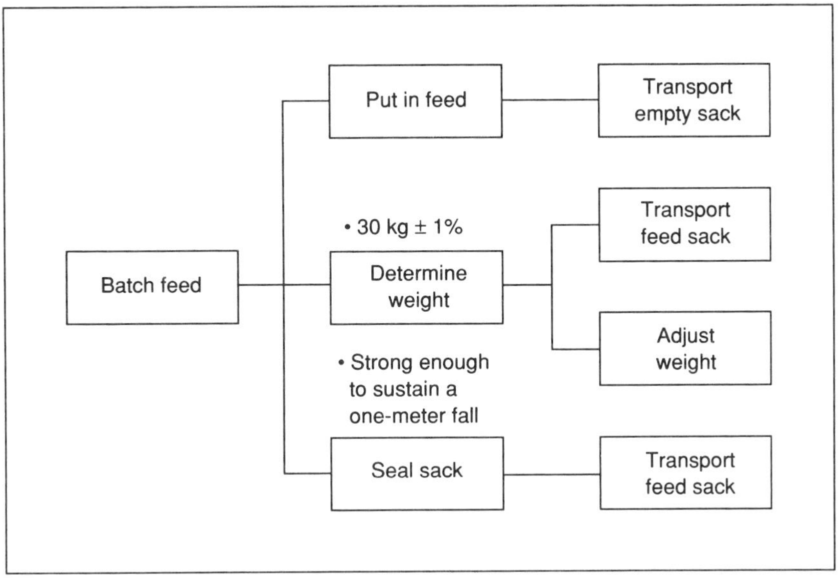

Figure 6-30. Function Family Tree for Feed-packing Process

Step 1: Understand the Object of Analysis

Gather information about the object of analysis. Process design, also called process planning, means replanning and launching construction of a series of processes in the form of a production system capable of guaranteeing quality, cost, and delivery.

Process design is generally based on product drawings and product specifications. In many instances, however, product drawings and specifications may not be adequate for the fabrication process. In such cases, product designs and specifications need to be reviewed from a manufacturing perspective.

In addition, process design involves not merely thinking about a way to build a product according to drawings; it requires consideration of a host of other critical factors, including

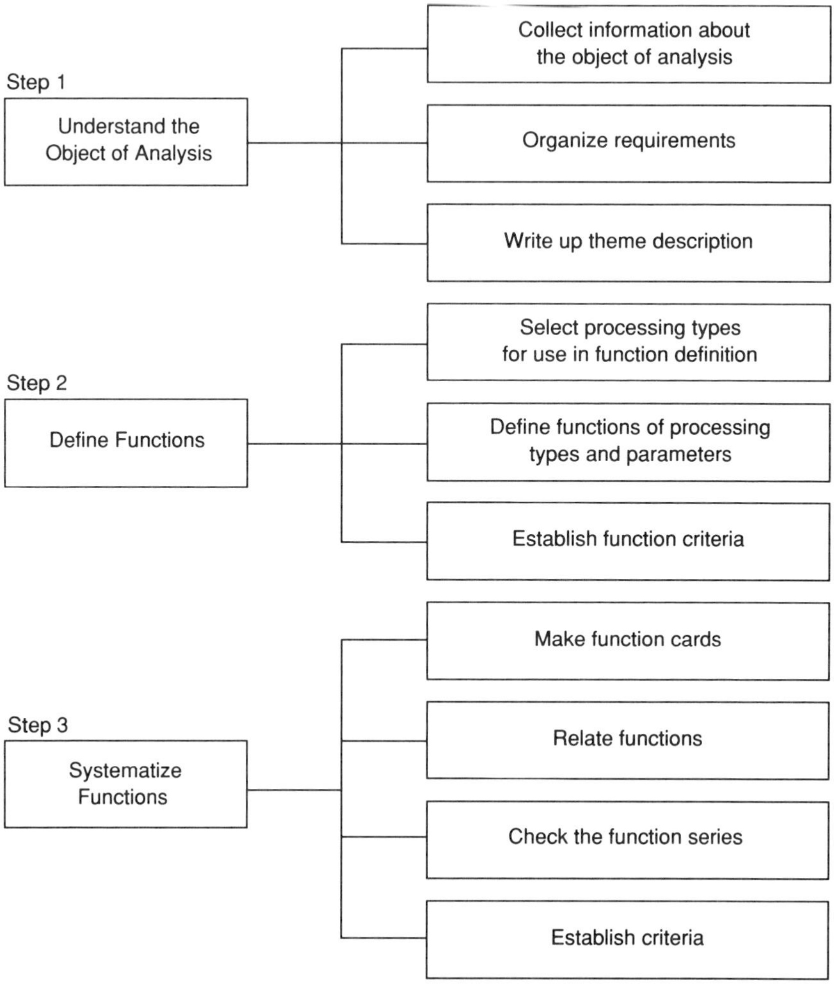

Figure 6-31. Process Design and Function Analysis Procedure

production volume, production method, production equipment, production costs and investment figures, and so forth.

While it is, of course, important to check product drawings and specifications sheets, applying function analysis to process design also demands collecting information relevant to process design and fully understanding what is required of the process to be designed.

Applying function analysis to process design begins with assembling requirements in the form of information such as that described in Table 6-7.

Table 6-7. Information to Be Collected for Process Design Requirements

Process design requirements relating to product specifications
- Specifications relating to product performance, reliability, maintainability, safety, and operability
- Product drawings (of overall structure, design, assembly, etc.)
- Parts drawings (materials, shapes, dimensions, tolerances, etc.)
- Quality-related demands, etc.

Process design requirements relating to production methods
- Production volume (planned no. of units, monthly no. of units, seasonal variation, etc.)
- Production system (batch production, continuous production, single line, mixed lines, etc.)
- Production equipment (equipment to be used, shared equipment, etc.)
- Production staffing (number of people, skill levels, etc.)

Process design requirements relating to production costs
- Target production costs (materials costs, processing costs, management costs, etc.)
- Target labor costs
- Allowable investment costs (equipment, tools and dies, etc.)

Other process design requirements
- Requirements concerning in-house vs. external fabrication
- Requirements concerning legal regulations
- Requirements concerning specifications and standardization

Organize the requirements. After they are organized, the various process design requirements collected in the form of information can be integrated into the design of the process. Not all requirements need be incorporated into the process design, however. Decide on final process design requirements (that is, all demands and conditions placed on the process design) after checking the appropriateness and conformity of factors such as specifications, production methods, and production costs. (See Figure 6-32.)

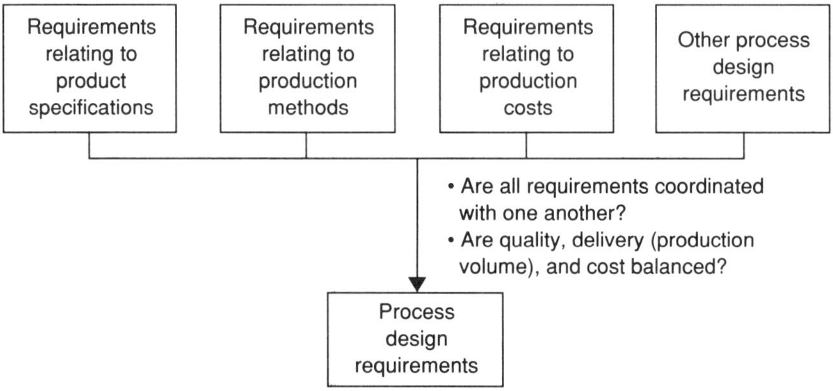

Figure 6-32. Establishing Process Design Requirements

Once all process design requirements have been determined, draw up a requirements list like the one shown in Figure 6-33.

Write out a theme description. When you have settled on process design requirements, write out a process design theme description based on those requirements. (See Figure 6-34.) A prerequisite to process design, such a description will incorporate information corresponding to a development specifications sheet.

Step 2: Define Functions

Select the processing types for use in function definition. The purpose of a process function is to transform objects (mate-

Function Analysis Worksheet 1

Requirements List
Product specifications requirements
• Drawings..
• Product quality demanded in fabrication ..
Production method requirements
• Production volume.................................
• Production system.................................
• Production space..................................
• Production equipment............................
• Production personnel.............................
Production cost requirements
• Target costs...
(etc.)

Figure 6-33. Process Design Requirements List

rials). Identifying that function is the starting point of process design. On the basis of previously identified process design requirements, we then proceed to check the processing types and parameters and so forth, and define the functions the process needs to carry out. (See Figure 6-35.)

Process design requirements include a variety of factors or conditions the process (that is, the production system) must satisfy. As Table 6-8 shows, each process design requirement is characterized by certain process-related factors.

Define the functions of processing types and parameters. With verb-noun pairs, define the functions the process is to

Function Analysis Sheet 2

Process Design Theme Description

1. Process name (production system name)

2. Technical demands and parameters

   - Drawings (see separate sheets, design drawings)
   - Quality demanded in fabrication (see separate sheets, design drawings)
   - Production system and production volume
   - Equipment and space used
   - Production personnel

3. Economic demands and parameters (from "requirements")

   - Target fabrication cost (per unit)
   - Target labor cost (per unit)
   - Allowable investment costs (equipment, tools, and dies)

4. Other process design-related demands and parameters (from "requirements")

   - In-house vs. external fabrication parameters
   - Legal regulations parameters
   - Specifications and standardization parameters

Figure 6-34. Process Design Theme Description

Function Analysis Worksheet 3

Requirement	Processing Types and Parameters
Product drawings	• List specific processing parameters based on product drawings
Product quality demanded in fabrication	• List specific processing types and parameters based on required quality

Figure 6-35. List of Processing Types and Parameters for Function Definition

Table 6-8. Characteristics of Process Design Requirements

Type	Process-related Characteristic
Product specifications requirements	• Requirements that indicate *what* is to be made (product) and *how* it is to be made (process) (These requirements are the main goal of process design and a prerequisite to function definition.)
Production method requirements	• Requirements involving production methods and technical factors (These requirements constrain processing methods serving as a production system.)
Production cost requirements	• Requirements relating to economic factors in production (These requirements constrain processing methods or may be seen as process design targets.)
Other requirements	• Requirements serving as technical constraints on production methods (These requirements constrain processing methods serving as a production system.)

perform by using the processing types and parameters you have checked for function definition.

Processing types and parameters listed on the basis of product drawings and required quality are expressed in narrative form, so it is important to recheck them and accurately interpret their process functions. (See Figure 6-36.)

Establish the function criteria. All defined functions are subject to various constraints on the ways and degrees to which they can be achieved. The next step is to determine those constraining conditions — or criteria — and relate them to each function. In terms of the various requirements we have already classified, we base such criteria either on production method requirements that limit the scope or means of function

Function Analysis Worksheet 4

Processing Types and Parameters	Functions		
	Verb	Noun	Criteria
(Processing types) • Put feed in sack and batch in uniform amounts. • Close sack securely without spilling feed.	put in batch close	feed feed feed sack	
(Processing parameters) • Guarantee that feed for each sack weights 30 kg ± 1% • Feed sacks do not break open even when dropped from a height of one meter	determine seal	weight sack	30 kg ± 1% resists one-meter falls

Figure 6-36. Defining Functions of Process Types and Parameters

Tips for Defining Functions

- Gain an adequate understanding of processing types and parameters.
- Treat each processing type and parameter as a function the process needs to perform.
- Recognize the limits of the functions to be performed.
- Use additional terms when it is difficult to define a function using one one verb and one noun.

(other points are the same as for process improvement)

achievement, or on processing parameters identified by product drawings or required quality.

In function analysis, we establish, or specify, such criteria in relation to functions expressed by verb-noun expressions.

A key question when relating individual criteria to functions is: Which criteria constrain the achievement of which functions? (See Figure 6-37.)

Criterion	Corresponding Function
Feed weight of 30 kg ± 1%	30 kg ± 1% constrains the weight of feed and so constrains the "determine weight" function.
Feed sacks resistant to one-meter drop	A one-meter drop means that the sack will not break even if dropped from a height of one meter. Sack strength is determined by product specification, so in terms of the process this criterion constrains the "close sack" function.

Figure 6-37. Relating Criteria to Functions

Process function criteria, ordinarily referred to as processing parameters or manufacturing specifications, serve as quality standards in the product manufacturing process. The more detailed and rigorous such quality standards are, the greater the constraint on achieving process function and the greater the cost of function performance. Therefore, you should designate only the minimum criteria needed.

In addition, you need to go back and recheck requirements or manufacturing specifications whenever criteria are unclear.

Step 3: Systematize Functions

Make function cards. Transfer each function identified by function definition onto a card like the one shown in Figure 6-38, so that you have the same number of cards as identified functions.

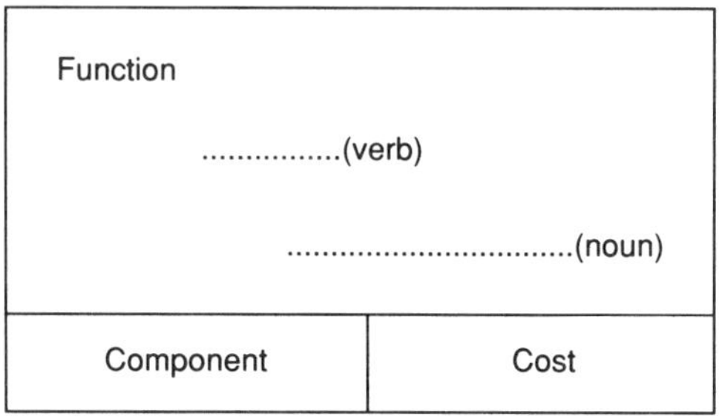

Figure 6-38. Function Card

In the space labeled "function" on each card, record the defined function in large block letters. For process design, leave the spaces labeled "component" and "cost" blank.

Relate the functions. Select a function card at random and look for the goal of that function by directing the question *why* to the function on the card. Identify the function matching the response to that question and place its function card to the left of the first card. (See Figure 6-39.)

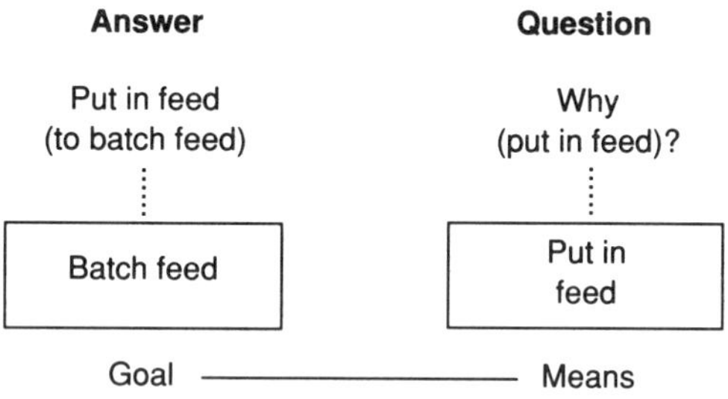

Figure 6-39. Relating Functions

If the "response" function is recorded on one of the cards you have already made, simply lay it down on the left. If it is not recorded, make up a new card and place it to the left of the "question" function card. You systematize functions by applying the above procedure for seeking goal functions to all the function cards. If the process just described fails to produce clear goal functions, revise your function definitions.

Verify the function series. When all defined functions have been positioned in the function series, verify the appropriateness of the series. Apply the following questions in checking the function series:

- Can the goal functions be reliably satisfied by the means functions?
- Are the means functions unnecessary if the goal functions are unnecessary?
- Are basic functions the process must perform incorporated without "leaks"?
- Can process design concepts be easily understood from the function series?

Functions should be added or corrected — or the function series revised — if either (a) means functions cannot reliably satisfy goal functions or (b) means functions serve any but necessary goal functions. When dealing with series of process functions, watch for the tendency to systematize functions by temporal relationships rather than by goal-means relationships.

Establish the criteria. On the completed function family tree, write in criteria corresponding to all the functions. In addition, write in criteria derived from production method requirements above corresponding functions on the function family tree. (See Figure 6-40.)

SUPPORT ACTIVITY IMPROVEMENT

The procedure for support activity improvement through function analysis is outlined in Figure 6-41. A more detailed examination of the process follows.

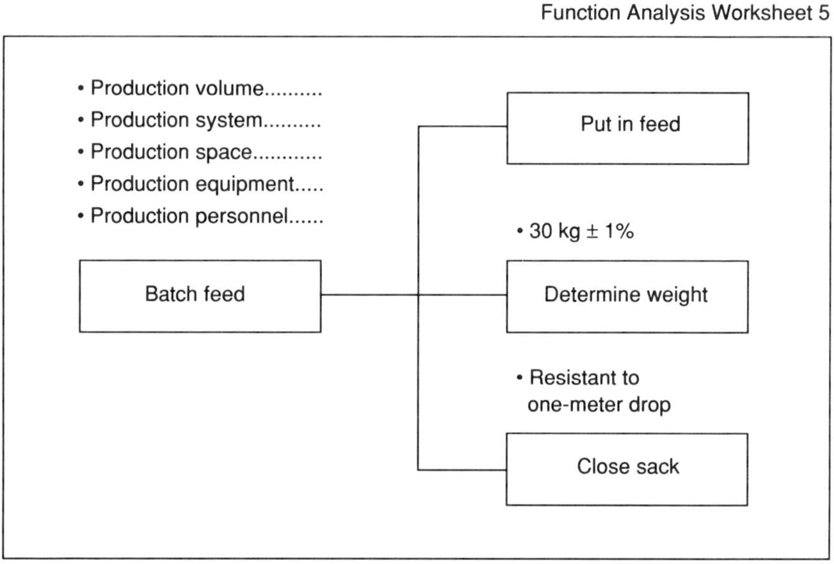

Figure 6-40. Establishing Criteria

Step 1: Select a Theme

Choose a support activity improvement theme. Support activity improvement means modifying current support activity methods to obtain better performance in function or cost.

The first step in applying function analysis to support activity improvement is theme selection: What support activity needs to be improved?

Support activities consist of mental tasks that involve little repetition. Such activities are characteristically difficult to improve: The conduct of work is not visible and evaluative measures are vague. Nevertheless, support activities harbor a variety of problems, such as quality, errors, and costs (which nearly always means labor costs). The support activity chosen as the target of function analysis should suffer from numerous such problems.

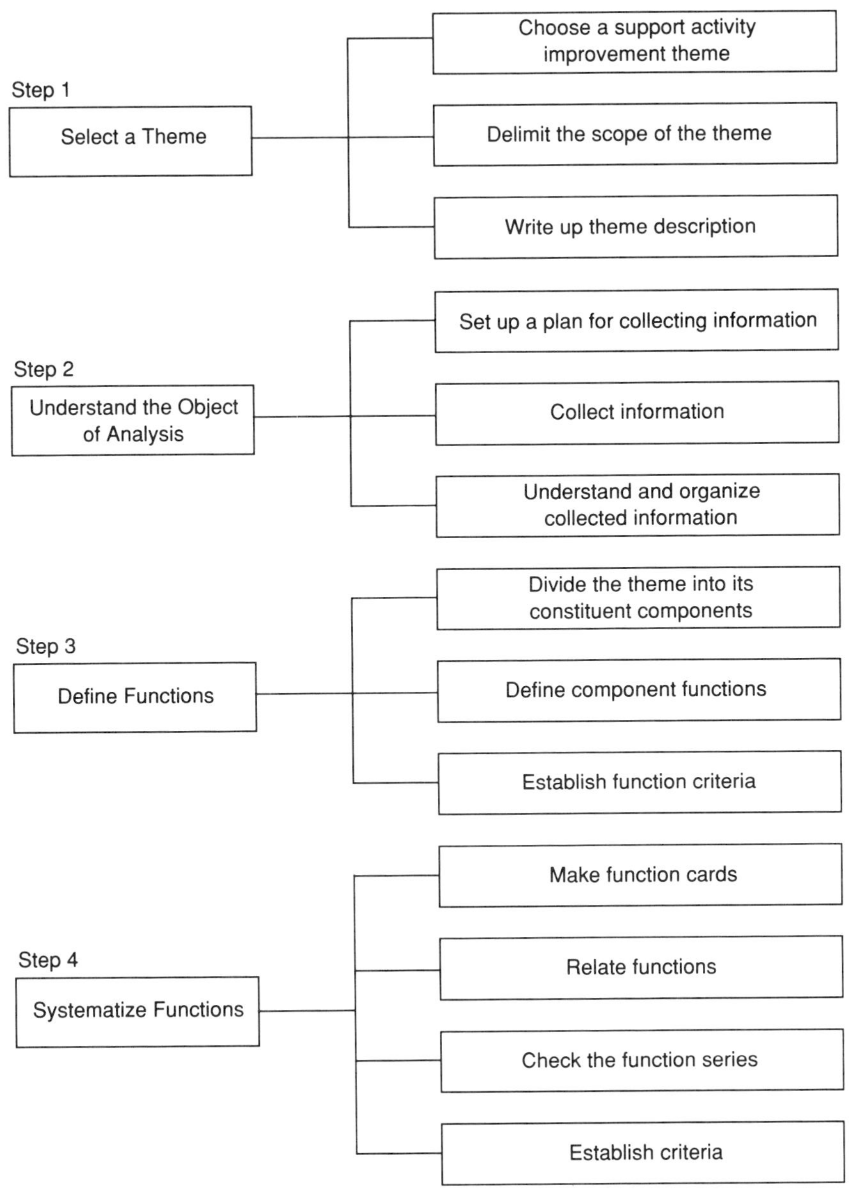

Figure 6-41. Support Activity Improvement and Function Analysis Procedure

The selection of a theme for support activity improvement should be carried out realistically, in accord with company circumstances. The points shown on the following checklist should be taken into account in this selection process.

- Are the goals and functions of the support activity clear?
- Are there any problems with the precision quality of the support activity?
- Does the support activity suffer from breakdowns or errors?
- Does the support activity suffer from delivery or timing problems?
- Does the support activity cost too much?
- Is the support activity too much trouble to carry out?
- Does the support system use the same methods it always has?
- Are there few constraints on changes in support activity methods?

Delimit the scope of the theme. The theme chosen for support activity improvement may be one of a number of different types. Support activity improvement rarely involves targeting all the support activities of a given organization — that is, all indirect and general affairs tasks. In the usual case, one specific problem-ridden activity will be the focus of improvement.

Limiting the theme to a specific activity makes it possible to carry out effective improvement of support work.

Once you have chosen a support activity improvement theme, you need to narrow its scope to clearly identify the domain in which function analysis will be used to make improvements. Limiting the scope of support activity improvement is especially important because support systems themselves are ill defined.

Write out a theme description. When you have chosen a support activity improvement theme and limited its scope, write out a one-page summary description of the theme. (See Figure 6-42.)

Function Analysis Worksheet 1

Description of Support Activity Improvement Theme

1. Summary and scope of support activity improvement theme
   - Summary of support activity improvement theme (What support activity is to be improved?)
   - Scope of support activity improvement theme (What does the theme support activity cover?)
2. Reasons for selecting support activity improvement theme (Why was this support activity chosen as a theme?)
   -
   -
   -
   -
   -
   -
3. Support activity improvement theme objectives
   - Quality improvement objectives (How much better do we want to make quality?)
   - Cost improvement objectives (How far do we want to lower costs?)
   - Other improvement objectives (In what other ways do we want to improve support activities?)

- "Summary and scope of support activity improvement theme" — Write down what the theme is and what it encompasses. Use language that is simple, specific, and easy to understand.
- "Reasons for selecting support activity improvement theme" — Briefly list each reason for choosing the theme.
- "Support activity improvement theme objectives" — For quality-related objectives, use numerical targets for numbers of errors, accuracy, timing, etc. For cost-related objectives, write in numbers of support activity personnel, labor costs, etc.

Figure 6-42. Description of Support Activity Improvement Theme

Step 2: Understand the Object of Analysis

Set up a plan for collecting information. Once you have settled on a process improvement theme, the next step is to understand the object of analysis by collecting information relevant to the theme. "Information" means knowledge needed for accomplishing various goals. In the case of support activity improvement, you need information on issues such as current support activity methods, the labor costs of individual support tasks, and current support activity problems.

Support activities characteristically involve mental tasks whose state and progress are invisible. The difficulty of subjectively perceiving and understanding support tasks themselves, in other words, makes it hard to improve such tasks. This makes it all the more important to collect relevant information and understand the object of analysis.

Setting up a plan for information collecting involves, first, identifying what information is needed, and then selecting information sources, a timetable, collection methods, and individuals with specific responsibilities.

Table 6-9 lists the types of information that should be collected with respect to a support activity improvement theme. Information is assembled using forms for information to be collected (see Figure 6-43) and an information collection plan (see Figure 6-44).

Collect information. Information-collecting activities, carried out on the basis of the information collection plan, generally take time and effort, so adherence to the plan is essential. Keep the following points in mind when collecting information:

- Information is knowledge needed to attain a goal. Keep the goal in mind when collecting information.
- Information is sometimes tainted by conjecture or prejudice. Collect the facts.
- Information is intangible. Record clearly each piece of information you collect.

Table 6-9. Information to Be Collected for a Support Activity Improvement Theme

Information relating to support activity requirements

- Support activity goals and objectives
- Criteria for improving support activities

Information concerning the nature of support activities

- Regulations concerning support activities
- Support activity flows
- Support activity procedures
- Support activity records

Information concerning support activity costs

- Breakdown of support activity labor costs
- Breakdown of other necessary support activity costs

Problems concerning support activities

- Incidence and types of support activity errors
- Incidence and types of support activity delays
- Wishes concerning support activity types and methods

- Criteria may be fictitious. Be sure to grasp only true criteria.
- A variety of problems will be encountered. Be sure to grasp problems accurately.

Understand and organize the collected information. All information collected provides important material for function analysis, and this makes it essential that information collected be fully understood. Inadequately understood is useless. A full understanding of the information also means an understanding of the product improvement theme that constitutes the object of analysis.

Collected information should be organized for ready access and ease of use in subsequent analyses.

Function Analysis Worksheet 2

Information to Be Collected

1. Information concerning support activity requirements (What is demanded of support activities?)
   -
   -
   -
2. Information concerning types of support activities (What do current support activities consist of?)
   -
   -
   -
3. Information concerning support activity costs (What do support activity costs consist of?)
   -
   -
   -
4. Problems concerning support activities (What are current support activity problems?)
   -
   -
   -

- "Information concerning support activity requirements" — List information concerning basic support activity goals and objectives, as well as conditions that constrain the execution of support activities.
- "Information concerning types of support activities " — List information for understanding the current composition of support activities.
- "Information concerning support activity costs " — List information for understanding the current state of costs incurred by support activities.
- "Problems concerning support activities" — List information for understanding problems concerning support activities.

Figure 6-43. Form for Collecting Information

Function Analysis Worksheet 3

Information Collection Plan					
Information Type	Information to Be Collected	Source(s)	Collection Schedule	Collection Method	Person in Charge
1. Information concerning support activity requirements	• • •				
2. Information concerning types of support activities	• • •				
3. Information concerning support activity costs	• • •				
4. Problems concerning support activities	• • •				

- "Information to Be Collected" — Transfer information to be collected from Function Analysis Worksheet 2
- "Source(s)" — Determine the appropriate departments or persons from whom each type of information is to be collected and write them in as sources.
- "Collection Schedule" — Set deadlines for the collection of each type of information and write in the corresponding date.
- "Collection Method" — Decide how each type of information is to be collected and write in the collection method.
- "Person in Charge" — Decide who is going to collect each type of information and write in that person's name.

Figure 6-44. Information Collection Plan

Step 3: Define Functions

Divide the theme into its constituent components. A support activity improvement theme corresponds to a specific system of support tasks. A support system will have various functions it needs to perform. To make it easier to define the various functions that must be performed, divide the support activity improvement theme into its constituent components and then define the function of each component.

There are various perspectives and units by which a support system may be broken down. The particular perspective and units by which we define the functions of constituent components will depend on the goals of support activity improvement. Figure 6-45 and Table 6-10 show how perspectives and units are affected by improvement goals.

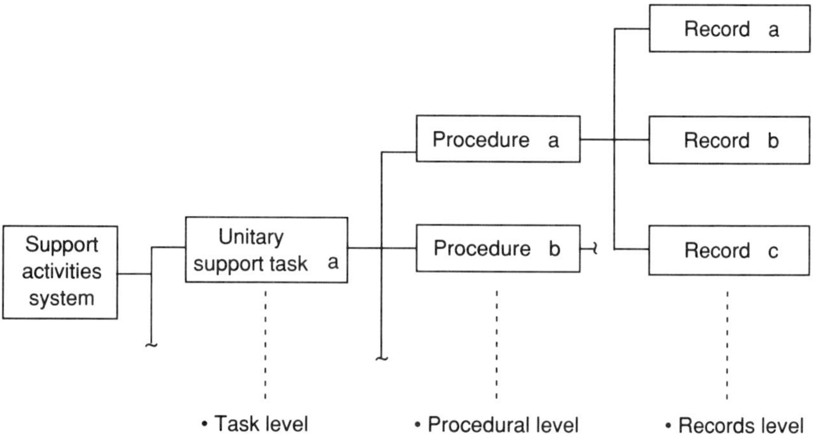

Figure 6-45. Support System Breakdown Structure

Define the component functions. Using verb-noun combinations, define the functions to be performed by the components you have identified.

In the case of a support activity improvement theme, functions to be performed by constituent components will be carried out either by people or organizational units and will consist of, for example, collecting or processing information. Other functions may involve decision making. Such components are not

Table 6-10. Support Activity Improvement Goals and Breakdown Structure

Improvement goals	Units	Approach to Analytic Units
System improvement	Task level	Function definition focuses on unitary tasks that make up a support system.
Task improvement	Procedural level	Function definition focuses on procedures that make up unitary support tasks.
Procedural improvement	Records level	Function definition focuses on records that make up support procedures.

visible, and different people will have different perceptions of what their proper functions are.

All this means that when defining component functions, you need to give thorough consideration to the purposes and essential roles of unitary support tasks, procedures, and records. You then define those purposes and roles as functions. (See Figure 6-46.)

Tips for Defining Functions

- Add terms to verb-noun expressions if necessary to make functions easier to understand.
- Avoid business jargon; use everyday language not subject to misinterpretation.
- Define functions without referring to the name of any specific business.
- Define functions using transitive verbs (i.e., verbs that take direct objects).
- Verify the purposes and roles of support activities and define them as functions.
- Keep the goals of the system as a whole in mind while treating the purposes and roles of support activities as functions.

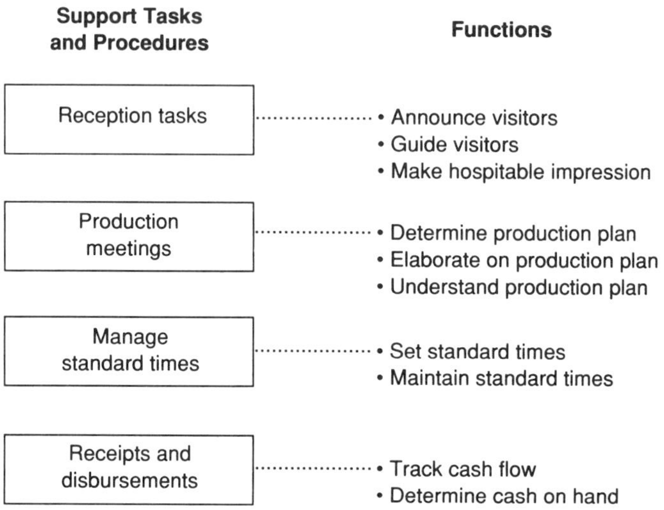

Figure 6-46. Functions as Purposes and Roles

Establish the function criteria. All defined functions are subject to various constraints on the ways and degrees to which they can be achieved, and the next step is to determine those constraining conditions — or criteria — and relate them to each function. In the case of support activities, criteria generally correspond to constraints on the quality of function achievement as measured by precision, accuracy, deadline performance, or time. Rarely are such factors clearly identified, however. Support activity factors corresponding to product quality or specifications, in other words, tend to be vague. This makes it crucial to use the information collected in step 2 to determine and establish criteria at appropriate levels from the point of view of the support system.

These criteria are a prerequisite to improving support activities by establishing alternative methods of performing support tasks. Strict criteria limit the scope of support activity improvement, whereas lax criteria facilitate such improvement. On the other hand, while narrow criteria make it easier to assess improvement, liberal criteria are vague and thus difficult.

Keep the above issues in mind when you establish criteria for functions. (See Figure 6-47.)

Function Analysis Worksheet 4

Support Task	Functions		
	Verb	Noun	Criteria

Figure 6-47. Establishing Function Criteria

Step 4: Systematize Functions

Make function cards. Transfer each function identified by function definition onto a card like the one shown in Figure

6-48, so that you have the same number of cards as identified functions.

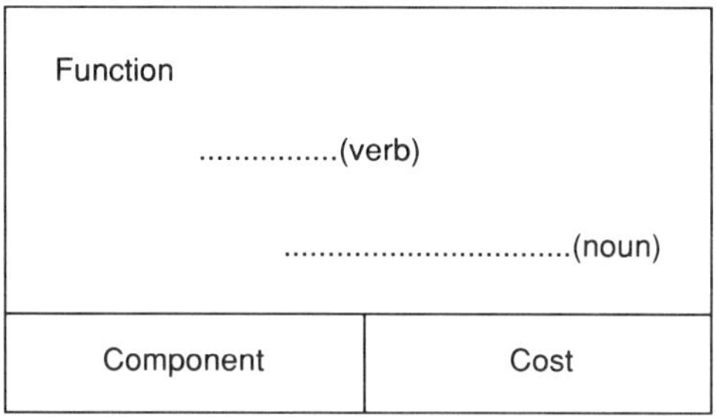

Figure 6-48. Function Card

In the space labeled "function" on each card, record the defined function in large block letters. In the space labeled "component," write the name or number of the support task or procedure whose function is defined.

Relate the functions. Select a function card at random and look for the goal of that function by directing the question *why* to the function on the card. Identify the function matching the response to that question and place its function card to the left of the first card. (See Figure 6-49.)

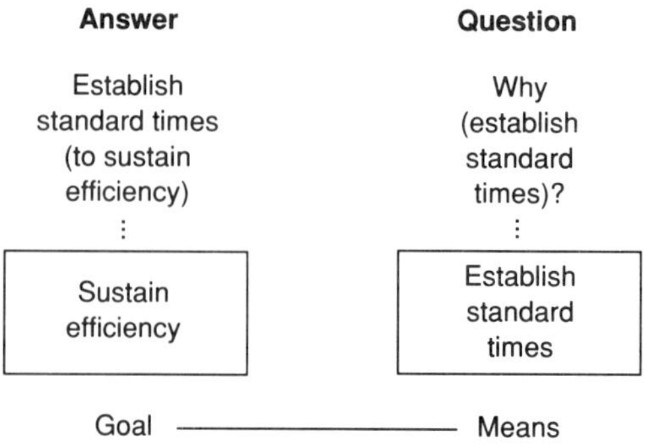

Figure 6-49. Relating Functions

If the goal function is not among already existing cards, make up a new card for it and place it to the left of the means function card. Continue searching for goal functions by repeating this procedure until a "response" function falls outside the scope of the support activities system theme. You systematize functions by carrying out the above procedure with all the function cards.

In analyzing support activity improvement themes, there is a tendency to treat the relationship between goals and means as a temporal one; in other words, to think of the existing task flow as reflecting goal-means relationships. When analyzing functions, it is crucial to think of goal functions straightforwardly, without regard for the current sequence of support tasks.

Check the function series. When all defined functions have been positioned in the function series, verify the appropriateness of the series. Apply the following questions in checking the function series:

- Can the goal functions be reliably satisfied by the means functions?
- Are the means functions unnecessary if the goal functions are unnecessary?

Means functions are inadequate — they "leak" — if they cannot reliably satisfy goal functions. A function needed for another function is incorrectly positioned, moreover, if its ostensible goal function is unnecessary. Either case calls for adding functions, changing the position of functions, or revising the function series.

Establish the criteria. Write in the corresponding criteria above the functions on the completed function family tree. Copy criteria set in the function definition stage above the functions they constrain. Establish and record new high-order goal function criteria where needed. (See Figure 6-50.)

Function Analysis Worksheet 5

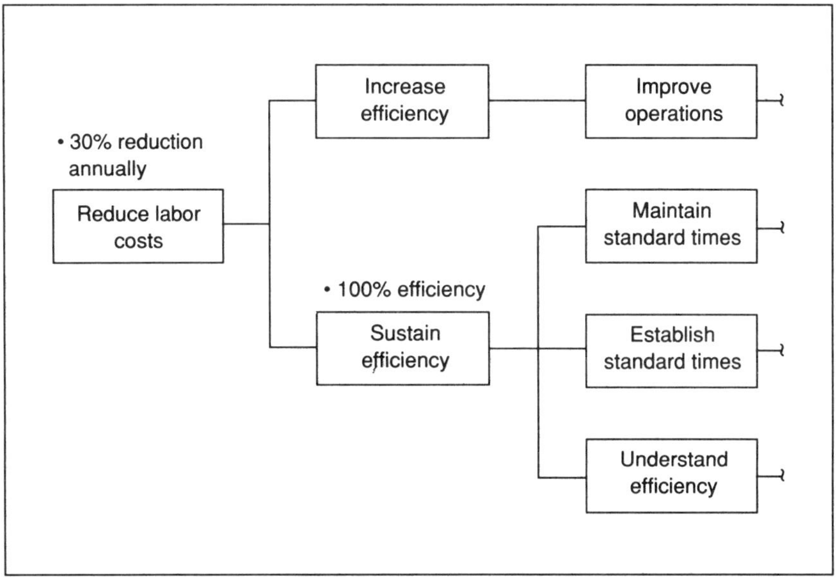

Figure 6-50. Function Family Tree for "Efficiency Control" System

SUPPORT ACTIVITY DESIGN

The procedure for support activity design through function analysis is analyzed in Figure 6-51. A detailed examination of each step follows.

Step 1: Understand the Object of Analysis

Collect information about the object of analysis. Support activity design means replanning and describing a high-quality, low-cost support system appropriate to work goals.

Generally speaking, support activities are rarely conducted on the basis of clearly identified demands made on a support system. Methods used to design support systems, moreover, are not always explicit. Support activities are nearly always

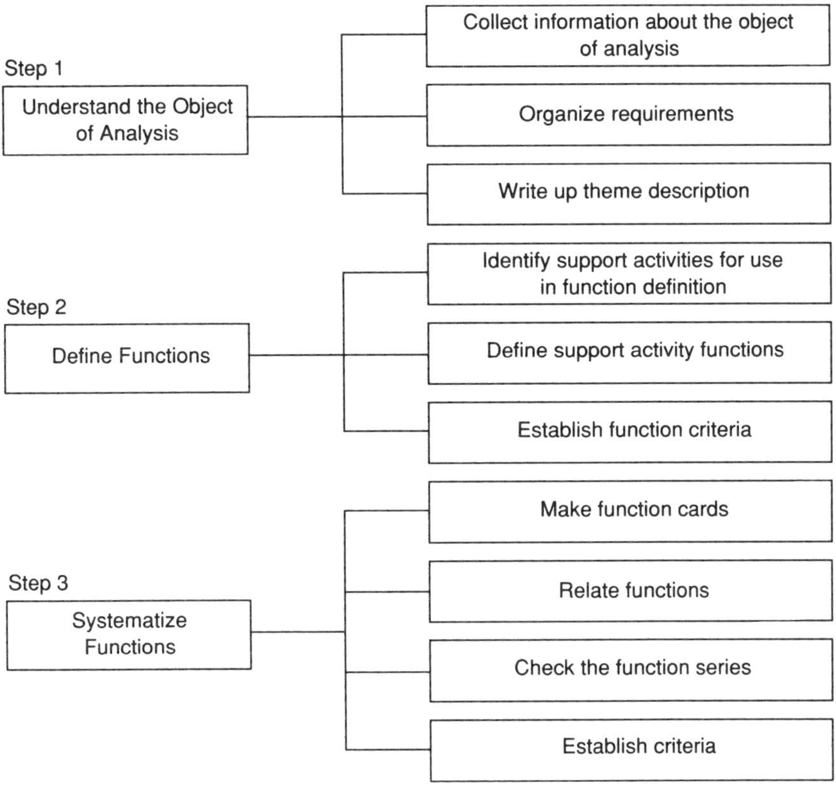

Figure 6-51. Support Activity Design and Function Analysis Procedure

designed by experience. Someone decides which organizational units will conduct the activities and then determines rules for carrying out the work.

The application of function analysis to the design of support activities begins with an accurate understanding of (1) what is required of the system to be designed, (2) what its goals are, and (3) what conditions should govern the design of the support system.

This makes it crucial to collect and understand information concerning the support system to be designed.

Applying function analysis to the design of support activities means, first, collecting information on requirements such as those shown in Table 6-11.

Table 6-11. Information to Be Collected for Support Activity Design

Support system requirements
• Goals and objectives of work a prerequisite to support system
• Requirements of each department with respect to support system
• Support system goals
• Support system objectives
Support system design requirements
• Rules and regulations affecting support system design
• Investment costs allowed for support system design
• Other support system design criteria, constraints
Support system administrative requirements
• Volume of support activities
• Time for support activities
• Precision of support activities
• Personnel and labor costs allowed by support activity administration
• Other support activity administrative criteria, constraints

Organize the design requirements. After they are organized, the various requirements collected in the form of information can be integrated into the design. Not all requirements need be incorporated into the design, however. Decide on design requirements (that is, all demands and constraints placed on the design of support activities) after making trade-offs among mutually contradictory requirements. (See Figure 6-52.)

Once all design requirements have been determined, draw up a requirements list like the one shown in Figure 6-53.

Write out a theme description. When you have settled on design requirements, write out a support activity theme description based on those requirements. (See Figure 6-54.) A prerequi-

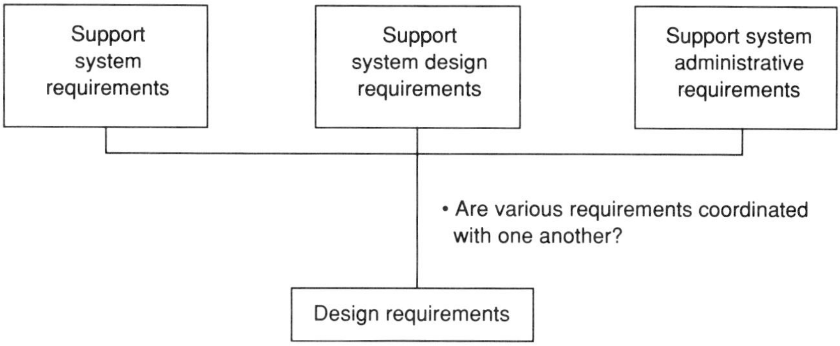

Figure 6-52. Establishing Support Activity Design Requirements

Function Analysis Worksheet 1

Requirements List

Support system requirements
- Work goals and objectives ....
- Departmental requirements ....
- Support system goals and objectives ....

Support system design requirements
- Rules and regulations ....
- Allowable design investment costs ....
- Other design criteria, constraints ....

Support system administrative requirements
- Volume and time of support activities ....
- Quality and precision of support activities ......
- Personnel for administration ....
- Other administrative criteria, constraints ....

Figure 6-53. Support Activity Design Requirements List

site to the design of support activities, such a description will incorporate information corresponding to a development specifications sheet for a product.

Function Analysis Worksheet 2

Support Activities Design Theme Description

1. Support system name

2. Support system requirements

   (a) Support activity goals and objectives
   (b) Departmental demands on support system
      • Department A
      • Department B
      • Department C

3. Support system design requirements

   (a) Rules and regulations
   (b) Allowable investment costs
   (c) Other

4. Support system administrative requirements

   (a) Volume and time of support activities
   (b) Quality and precision of support activities
   (c) Administrative personnel
   (d) Other

Figure 6-54. Support Activity Design Theme Description

Step 2: Define Functions

Identify the support activities for use in function definition. Functions are the goals and duties of the support activity to be designed. Support activity design begins with the identification of what those functions are.

Unlike engineering-oriented product functions, support activity functions take the form of purposes and roles to be performed by individuals or organizational units. Requirements underlying support activity functions, moreover, are discursive rather than technical. You need to look at each requirement and sort out the support activities that will fulfill functions demanded by (1) the goals and objectives of the overall support system and (2) individual departmental requirements. Support activities designed to fulfill both overall system goals and objectives and individual departmental requirements become a means of achieving support goals. At the same time, they are actions for carrying out individual tasks.

With support activities consisting of individual actions for achieving goals and objectives, it is crucial to use brainstorming techniques to generate ideas for such activities. It is important, too, to restrict the scope of activity possibilities by using multiple criteria for summary evaluations of diverse ideas.

Evaluate support activities as either to be adopted (√) or not (×). Activities to be adopted will then be used in the next step, function definition. (See Figure 6-55.)

Function Analysis Worksheet 3

Support system name: Efficiency Control System
Support system goals/objectives: Reduce labor costs in mfg by 30% each period

Support activities for achieving goals and objectives	Evaluation
• Cut labor costs by reducing processing work-hours	√
• Control efficiency by establishing standard times	√
• Cut processing labor costs by improving processing and work methods	√
• Survey efficiency to identify areas of wasted labor costs	√
• Adhere to standard times accurately and consistently	√
• Raise efficiency by guiding operators in their tasks	√
• Grasp efficiency accurately	√
• Accurately grasp performance times and outputs	√
• Conduct time studies	√
• Use *kaizen* techniques to improve work methods	√
• Study in order to make the use of *kaizen* techniques possible	×
• Carry out improvement activities	√

Figure 6-55. Generating Ideas for Function Definition

Define the support activity functions. Using verb-noun pairs, define functions to be performed by each support activity idea you have generated.

Although each support activity you have come up with may be the same as a function to be performed, it is not necessarily expressed in the form required by function analysis. Review each support activity, therefore, and then use a verb-noun combination to express and define its function. (See Figure 6-56.)

Function Analysis Worksheet 4

Support Activity	Function		
	Verb	**Noun**	**Criteria**
• Cut labor costs by reducing processing work-hours.	Reduce Reduce	work-hours labor costs	
• Control efficiency by establishing standard times.	Establish Control	standard times efficiency	
• Cut processing labor costs by improving processing and work methods.	Improve Reduce	tasks labor costs	
• Survey efficiency to identify areas of wasted labor costs.	Grasp Identify	efficiency labor cost waste	
• Adhere to standard times accurately and consistently.	Adhere to	standard times	
• Raise efficiency by leading operators in their tasks.	Lead Maintain	work efficiency	

(etc.)

Figure 6-56. Defining Functions of Support Activities

Tips for Defining Functions

- Add terms to verb-noun expressions if necessary to make functions easier to understand.
- Use everyday language not subject to misinterpretation.
- Define functions without referring to specific support activity methods.
- Verify the basic purposes and roles of support activities and define them as functions.
- Define functions using transitive verbs (i.e., verbs that take direct objects).
- Keep the goals of the system as a whole in mind while treating the purposes and roles of individual support activities as functions.

Establish the function criteria. All defined functions are subject to various constraints on the ways and degrees to which they can be achieved. Support systems typically have never encompassed any notion of function or anything like product "quality" or specifications. This makes support systems all the more vague.

Among previously identified requirements, criteria are constraining conditions that, for example, limit the means by which — or prescribe the degree to which — functions can be achieved. In function analysis, these criteria are established in relation to functions expressed as verb-noun combinations.

In relating criteria to functions, be sure to identify clearly each function whose achievement is constrained by a given criterion. (See Figure 6-57.)

Choose criteria from among requirements of the support system, support system design, and support system administration. Establish the criteria by checking them and then relating them to functions. Additional information gathering and verification is needed whenever criteria are unclear.

Criterion	Corresponding Function
• Support system objective Reduce labor costs by 30% each period	• The percentage reduction of labor costs constrains and corresponds to the "reduce labor cost" function
• Support system objective Reduce work hours by 30% each period	• The percentage reduction of work-hour reduction constrains and corresponds to the "reduce work-hour" function
• Departmental requirement Establish realistic standard times	• Establishing realistic standard times refers to the precision of standard times and constrains the "adhere to standard times" function

Figure 6-57. Relating Criteria to Functions

Step 3: Systematize Functions

Make function cards. Transfer each function identified by function definition onto a card like the one shown in Figure 6-58, so that you have the same number of cards as identified functions.

In the space labeled "function" on each card, record the defined function in large block letters. Leave the spaces labeled "component" and "cost" blank.

Relate the functions. Select a function card at random and look for the goal of that function by directing the question *why* to the function on the card. Identify the function matching the

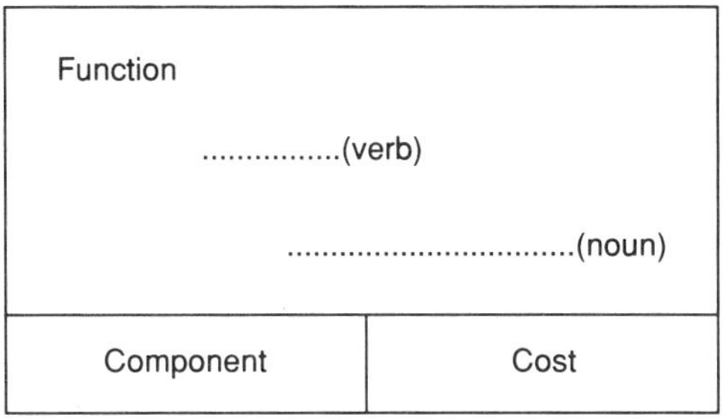

Figure 6-58. Function Card

response to that question and position its function card to the left of the first card. (See Figure 6-59.)

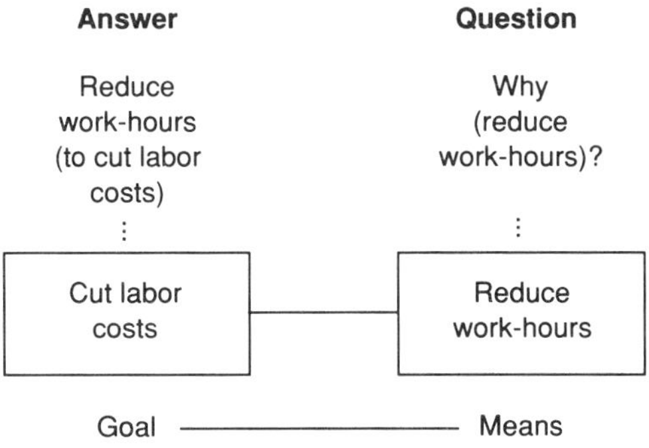

Figure 6-59. Relating Functions

If the "response" function is recorded on one of the cards you have already made, simply lay it down on the left. If it is not recorded, make up a new card and place it to the left of the "question" function card. Continue searching for goal functions by repeating this procedure until a "response" function falls outside the scope of the design target. Systematize the functions by carrying out the above procedure with all the function cards.

If the process just described fails to produce clear goal functions, revise your function definitions.

Check the function series. When all defined functions have been positioned in the function series, verify the appropriateness of the series. Apply the following questions in checking the function series:

- Can the goal functions be reliably satisfied by the means functions?
- Are the means functions unnecessary if the goal functions are unnecessary?
- Are basic functions the support activity must perform incorporated without "leaks"?
- Can support system design concepts be easily understood from the function series?

Functions should be added or corrected — or the function series revised — if either (1) means functions cannot reliably satisfy goal functions or (2) means functions serve any but necessary goal functions.

Functions or the function series should be revised, too, if the basic functions expected of the support system are not adequately accounted for or if support system concepts are not easily understandable.

Establish the criteria. On the completed function family tree, write in criteria corresponding to all the functions. (See Figure 6-60.) If all criteria established in the function definition stage are located on the function family tree, check to see that all the criteria originating in the design requirements figure on the chart as well.

QUALITY IMPROVEMENT

The procedure for quality improvement through function analysis is outlined in Figure 6-61. A more detailed examination of each step follows.

Function Analysis Worksheet 5

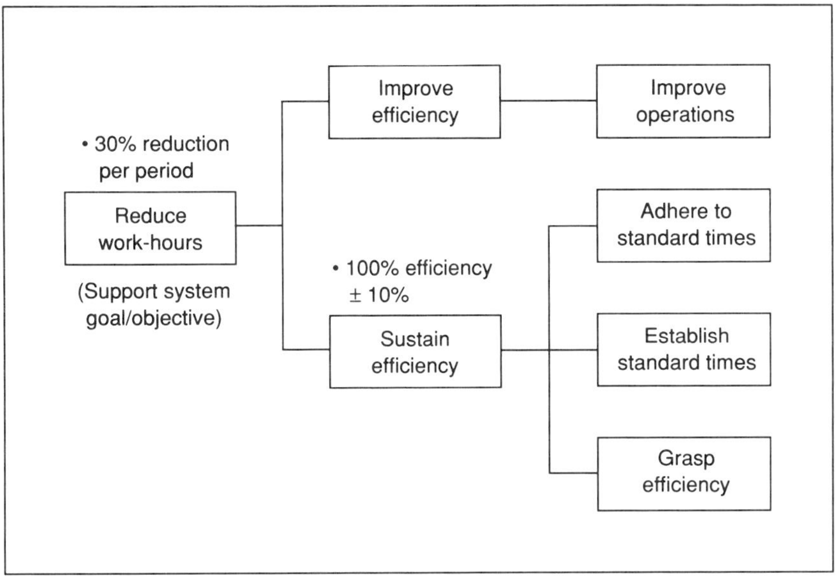

Figure 6-60. Establishing Criteria

Step 1: Select a Theme

Plan and carry out information collecting. Quality improvement means modifying a current production process or product merely to obtain better quality. Quality improvement has traditionally been carried out within the realm of QC and has addressed itself, in particular, to process management and control. By contrast, the kind of quality improvement we are referring to here focuses on design and is carried out via changes in the design of products and processes.

All products and processes harbor certain quality problems. A product or process chosen as the target of function analysis should suffer from numerous such quality problems.

Identifying a problem-ridden product or process is not as simple as choosing a theme for product improvement or process

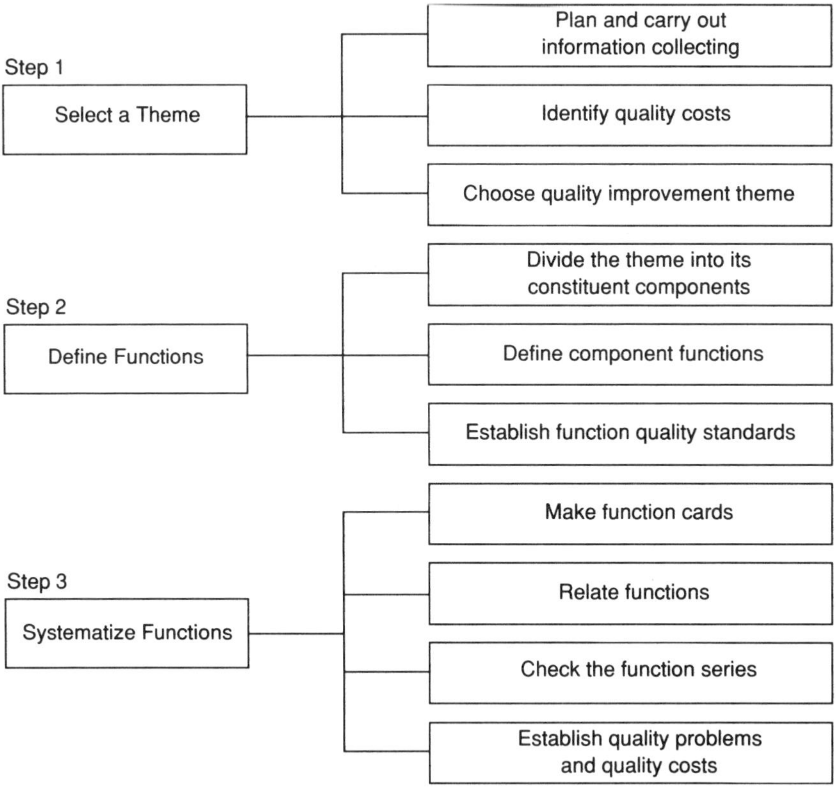

Figure 6-61. Quality Improvement and Function Analysis Procedure

improvement. This is because we must select problem-ridden products or processes indirectly, via a resulting quality problem. When we choose a product or process as a target for improvement, in other words, we need only settle on an independent product or process suffering from numerous problems. All too often, however, quality problems do not at first clearly implicate a problematic product or process. In such cases, selection of a target theme requires us to identify the essence of the problem by collecting extensive information on it. Information is knowledge and therefore not something you can pick up and hold in your hand. This makes it crucial to establish a plan and assemble information on quality problems in such a way that permits the accurate selection of a product or process that can serve as a product improvement theme.

Setting up a plan for information collecting involves, first, identifying what information is needed, and then selecting information sources, a timetable, collection methods, and individuals with specific responsibilities.

Table 6-12 lists the kind of information that should be collected with respect to a quality improvement theme. An information collection plan is carried out using forms for information to be collected (see Figure 6-62) and an information collection plan (see Figure 6-63).

Table 6-12. Information to Be Collected for a Quality Improvement Theme

Information on quality problems involving product use
- Irregularities and quality claims concerning product use
- Users' desires concerning product use
- Costs (losses) due to irregularities, and quality claims concerning product use

Information on quality problems involving product sales and distribution
- Irregularities and quality claims concerning product sales and distribution
- Desires concerning product sales and distribution
- Costs (losses) due to irregularities, and quality claims concerning product sales and distribution.

Information on quality problems involving product manufacture
- Data on rejects and yield in product manufacture
- Data on rework
- Precision of support activities
- The cost of quality in product manufacture (costs of control and losses)

Function Analysis Worksheet 1

Information to Be Collected

1. Information on quality problems involving product use
   •
   •
   •
2. Information on quality problems involving product sales and distribution
   •
   •
   •
3. Information on quality problems involving product manufacture
   •
   •
   •

Figure 6-62. Form for Collecting Information

Information-collecting activities, carried out on the basis of the information collection plan, generally take time and effort, so adherence to the plan is essential. Keep the following points in mind when collecting information:

- Information is knowledge needed to attain a goal. Keep the goal in mind when collecting information.
- Collect information in the form of facts.
- As much as possible, assemble information in a quantitative form.
- Organize information in a way everyone can understand.
- Quality may be fictitious. Be sure to grasp only true quality.
- Quality problems may be fictitious. Be sure to grasp only true quality problems.
- Collect information from accurate sources.
- Make effective use of QC techniques in organizing information on quality problems.

Function Analysis Worksheet 2

Information Collection Plan						
Information Type	Information to Be Collected	Source(s)	Collection Schedule	Collection Method	Organiza-tion Method	Person in Charge
1. Use Information	• • •					
2. Sales and Distribution Information	• • •					
3. Manufacturing Information	• • •					

- "Information to Be Collected" — Transfer information to be collected from Function Analysis Worksheet 1
- "Source(s)" — Determine the appropriate departments or persons from whom each type of information is to be collected and write them in as sources.
- "Collection Schedule" — Set deadlines for the collection of each type of information and write in the corresponding date.
- "Collection Method" — Decide how each type of information is to be collected and write in the collection method.
- "Person in Charge" — Decide who is going to collect each type of information and write in that person's name.

Figure 6-63. Information Collection Plan

All collected information related to quality problems provides important material for function analysis. This makes it essential that the information collected be fully understood and organized so it can be used whenever needed.

Identify the quality costs. Once all information relating to quality problems has been assembled, the next step is to use that information to select a product or process whose quality is to be improved. This product or process will become the quality improvement theme — in other words, you will choose a product or process as a quality improvement theme after identifying quality costs on the basis of the information you have collected. Costs guide the selection of a theme for two reasons: (1) quality problems originally appear as costs, and (2) costs can be used to measure the extent of quality problems.

There are three kinds of quality costs: reject costs, evaluation costs, and prevention costs. Table 6-13 gives details on each of these types.

Table 6-13. Quality Costs

• Reject Costs	Costs incurred because of nonconformity to product specifications (material and labor costs due to scrap, defects, rework, etc.)
• Evaluation Costs	Costs incurred because of inspections or sorting (labor costs of inspection, sorting, etc.)
• Prevention Costs	Costs incurred in order to prevent the occurrence of defects (labor costs of quality control, etc.)

Once the quality costs have been identified, classify each one by product or process to find the one that generates the highest quality costs.

Organize cost information using a form like the one shown in Figure 6-64. Clearly show what units you are using to express occurrence of quality costs (such as per product unit or per month).

Choose a quality improvement theme. Once you have identified quality costs by product and process, use that information to select the quality improvement theme. If a specific

Function Analysis Worksheet 3

Process / Product	Use	Distribution	Manufacturing			Total
			Process X	Process Y	Process Z	
Product A	REJ... EVAL... PREV...	REJ... EVAL... PREV...	REJ... EVAL... PREV...	REJ... EVAL... PREV...	REJ... EVAL... PREV...	REJ... EVAL... PREV...
Product B	REJ... EVAL... PREV...	REJ... EVAL... PREV...	REJ... EVAL... PREV...	REJ... EVAL... PREV...	REJ... EVAL... PREV...	REJ... EVAL... PREV...

- REF = Reject costs; EVAL = Evaluation costs; PREV = Prevention costs
- Fill in "Total" column only when product or process costs cannot be broken down.

Figure 6-64. Classifying Quality Costs by Product or Process

product or process is responsible for an inordinate proportion of quality costs, then that product or process becomes the target of quality improvement — or the theme.

When no one product or process generates a disproportionate share of the quality costs, select a group of products or series of processes as the object of process improvement. In either case, choose as a theme specific products or processes that are generating problems in terms of costs.

Constructing quality cost Pareto diagrams by product and process can make deliberations on the selection of a quality improvement theme more manageable.

Once a theme is selected, check the scope of the product (or products) or process (or processes) involved, identify the product or process, and write out the information on a suitable form. (See Figure 6-65.)

Step 2: Define Functions

Divide the theme into its constituent components. A quality improvement theme is a specific product or process that

Function Analysis Worksheet 4

Quality Improvement Theme Description

1. Quality improvement theme summary and scope

   (a) Summary of quality improvement theme (What is the theme?)

   (b) Scope of quality improvement theme (What is the scope of the theme?)

2. Reasons for choosing quality improvement theme (Why was this theme selected?)

   •

   •

   •

3. Quality objectives for theme

   (a) Quality-related objectives (How much do we want to improve quality?)

   (b) Cost-related objectives (How much do we want to lower costs?)

- "Quality improvement theme summary and scope" — state clearly and simply whether theme is product-related or process-related, as well as its scope.
- "Quality objectives for theme" — clearly state quality-related objectives in terms of numerical characteristic values, and cost-related objectives in terms of the rate and amount of cost reduction.

Figure 6-65. Quality Improvement Theme Description

generates quality costs. The fact that the product or process involves high costs means that problems lie either in product design (if the theme is a product) or (if the theme is a process) in processing methods — in other words, in process design. You use function analysis to determine what the essential function of the product or process in question is. Then, by establishing a different method for satisfying that function, you can redesign the product or process so that no quality costs are generated. (See Figure 6-66.)

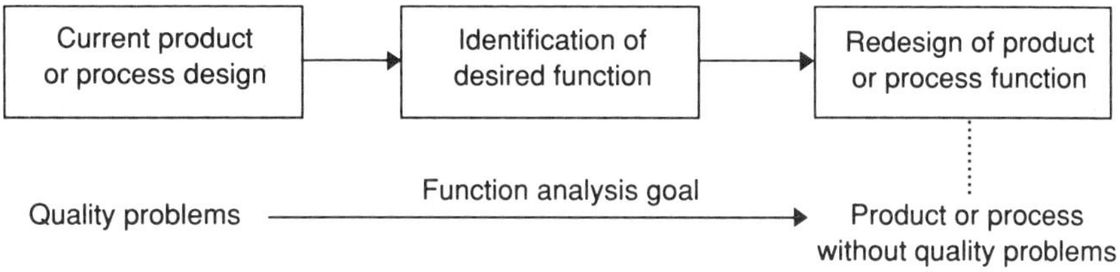

Figure 6-66. Defining Function Analysis Goal

To define functions, you must first identify the function (or functions) the product or process is supposed to perform. A product or process will have a number of such functions. Identifying those functions involves breaking the product or process down into constituent components and then defining the function of each component. For both products and processes, define the functions of components that correspond to the level of the quality problem.

Table 6-14 outlines an approach to identifying constituent components.

Table 6-14. Division into Components

Quality Problem / Object of Improvement	When a basic and significant quality problem prevails	When numerous small quality problems prevail
Product or component part	Treat large assembly units as components	Treat parts as components
Processes or series of processes	Treat individual processes or single operations as components	Treat individual operations or motions as components

There is no set limit on the number of components into which a product or process may be broken down, but it is preferable to have between ten and 30 components.

Define the component functions. Using verb-noun combinations, define the functions to be performed by the components you have identified.

Ordinarily, a single constituent component will not perform a function on its own; rather, it will carry out a specific function as part of a product or process constituting a system. To define component functions, you define what function each constituent component should perform in the context of such a system. Stay alert to the possibility that a single component may have to fulfill multiple functions.

Tips for Defining Functions

- Define functions simply, with verb + noun expressions.
- Use easy language anyone can understand.
- Use specific terms not subject to misinterpretation.
- Define functions without referring to specific names of parts or operations.
- Avoid using proper nouns in definitions.
- Use transitive verbs (i.e., verbs that take direct objects).
- Use abstract function expressions whenever possible.
- Use easily quantifiable function expressions whenever possible.
- Use affirmative expressions rather than negative ones.

(*Quality expression:* "Don't generate noise") vs.
(*Function expression:* "Prevent noise")

Establish function quality standards. All defined functions are subject to various constraints on the ways and degrees to which they can be achieved. These functional constraints —

or criteria — originate in product or process specifications, and what is generally referred to as quality corresponds to the levels and variability of those specifications. Current function criteria — that is, quality — may not be suitable from the customer's perspective. Often, too, quality criteria may never have been made explicit.

As a result, quality criteria must be reestablished at this point. This means designating product or process quality standards as function criteria, thereby clearly identifying them as standards for quality improvement.

Indeed, current quality problems are simply cases in which the product or process does not conform to these quality standards or constraints. You therefore describe quality problems in concrete terms by relating functions to their criteria (again, quality). In this way, you identify quality standards and quality problems from the perspective of product or process function. (See Figure 6-67.)

Function Analysis Worksheet 5

No.	Constituent Component	Functions			Quality Problem
		Verb	Noun	Quality Standards	

Figure 6-67. Identifying Quality Standards and Quality Problems

Step 3: Systematize Functions

Make function cards. Transfer each function and quality standard identified by function definition onto a card like the one shown in Figure 6-68, so that you have the same number of cards as identified functions. In the space labeled "function" on each card, record the defined function in large block letters. In the space labeled "component," write the name or number of the constituent component whose function is defined. In the space labeled "quality standard," write the function criterion; that is, the quality criterion.

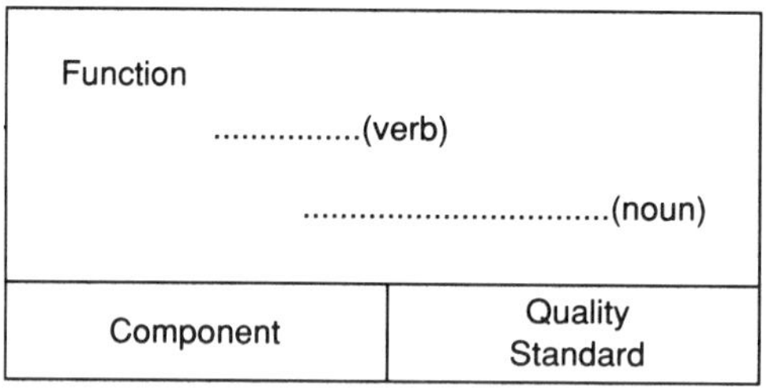

Figure 6-68. Function Card

Relate the functions. Select a function card at random and look for the goal of that function by directing the question *why* to the function on the card. Identify the function matching the response to that question and place its function card to the left of the first card. (See Figure 6-69.)

If the "response" function is recorded on one of the cards you have already made, simply lay it down on the left. If it is not recorded, make up a new card and place the new card to the left of the "question" function card. Continue searching for goal functions by repeating this procedure until a "response" function falls outside the scope of the improvement target. You systematize functions by carrying out the above procedure with all the function cards.

Object \ Q & A	Answer Perform in order to ..... (verb) some .....(noun)	Question Why perform the means function?
When the quality improvement theme is a product	Provide light Goal function	Send current Means function
When the quality improvement theme is a process	Batch feed Goal function	Put in feed Means function

Figure 6-69. Relating Functions

If the process just described fails to produce clear goal functions, revise your function definitions.

There is a tendency, when the quality improvement theme is a process, to treat the next operation or process as a goal function. Although the next operation or process may, indeed, be the goal function of the operation or process that precedes it, this is not so in every case. It is important to think of goal functions straightforwardly, without regard for the current task sequence or process series.

Check the function series. When all defined functions have been positioned in the function series, verify the appropriateness of the series. Apply the following questions in checking the function series:

- Can the goal functions be reliably satisfied by the means functions?
- Are the means functions unnecessary if the goal functions are unnecessary?

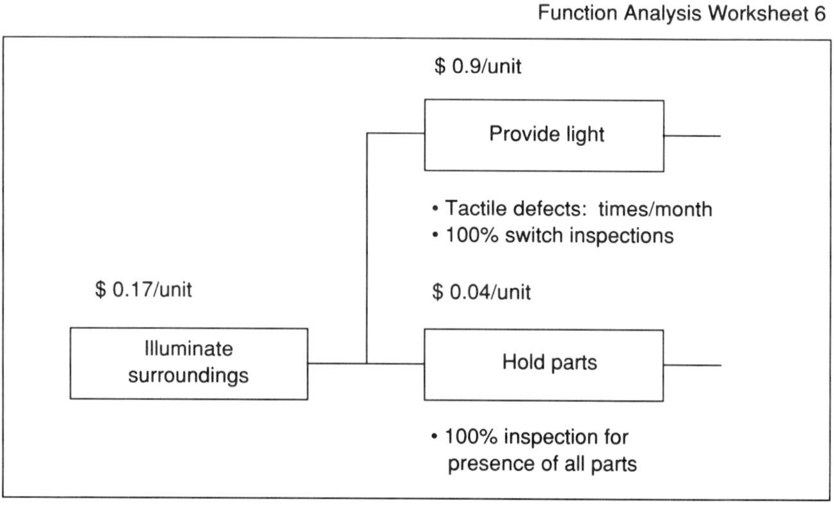

Figure 6-70. Example 1: Establishing Quality Problems and Quality Costs in a Pocket Flashlight

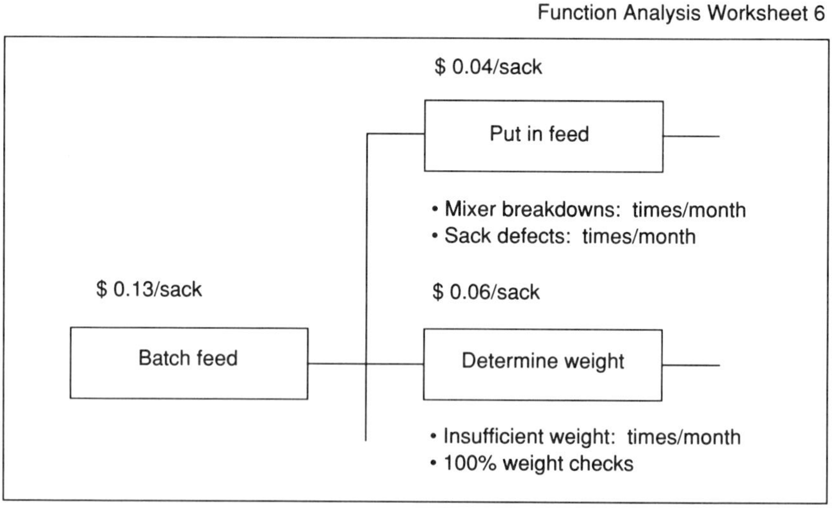

Figure 6-71. Example 2: Establishing Quality Problems and Quality Costs in a Feed-packing Process

Means functions are inadequate — they "leak" — if they cannot reliably satisfy goal functions. Revise the function series by changing the position of any function that is needed for another function but whose ostensible goal function is unnecessary.

In other cases, functions may be wanting from the point of view of customer demands or other issues. Here again, revise the function series by adding functions.

Establish quality problems and quality costs. Above individual functions on the completed function family tree, write in corresponding quality problems and the quality costs they generate.

Write specific quality problems below the functions where they occur and quality costs above the functions where they occur. (See Figures 6-70 and 6-71.)

Relating quality problems and costs to individual functions on a function family tree allows you to see at a glance which functions involve multiple problems and high costs. At the same time, the function family tree serves as a quality improvement guide by pointing to function attainment methods whose improvement can resolve product and process quality problems.

7
Function Analysis in Action

FUNCTION ANALYSIS AND SYSTEM UNDERSTANDING

The Need for System Understanding

We can make a distinction between two kinds of systems: hardware systems, such as products or equipment, and software systems, such as organizations or administrative work. Both types of systems (1) are made up of multiple constituent components, (2) possess mutual interactions and functions among those constituent components, and (3) have specific overall goals and functions.

What this means is that a system is composed of observable elements combined for the purpose of performing individual component functions or overall system functions.

When considering, say, product or service systems, people have a tendency to look at and understand only their immediately observable aspects. They judge products by examining their structure or form, while they judge services by looking at observable actions. People need to remember, however, that all systems — including products and services — exist in order to perform functions. The successful fulfillment of functions is what systems and their components are for.

Inadequate understanding of the essential functional nature of systems leads to the kinds of problems shown in Figure 7-1.

System	Problems Due to Inadequate System Understanding
Product system	"We don't understand the product's concept or functional characteristics"
	"Too much waste in product design"
	"We can't make fundamental changes in the product"
Production system	"We don't know what the goals of the processes or operations are"
	"Too much waste in processes and operations"
	"We can't make fundamental changes in the processes or operations"
Equipment system	"Equipment isn't being used correctly"
	"Too much waste in the equipment"
	"We can't make fundamental changes in the equipment"
Organizational system	"We don't understand the structure or role of the organization"
	"Too much waste in the organization"
	"We can't make fundamental changes in the organization"
Management and support system	"We don't understand the structure or goals of management and support activities"
	"Too much waste in management and support activities"
	"We can't make fundamental changes in management and support activities"

Figure 7-1. Problems Due to Inadequate System Understanding

Applying function analysis to a system lays a firm foundation for efficient and waste-free management by making the system's functional essence clear to everyone.

System Understanding and the Effects of Function Analysis

Function analysis is analysis of systems in terms of their functions. An analysis of their purposes and roles allows you to understand the essence of systems. This understanding helps you to resolve various problems arising from inadequate understanding of those systems.

Specifically, the use of function analysis to understand a product system:

- deepens your understanding of the product system
- allows everyone to understand the essence of the system by revealing product concepts and characteristics
- reveals the significance of product structure, materials, dimensions, and so on, and enables you to find waste

The use of function analysis to understand a production system:

- deepens your understanding of the production system, process, operation, and so on
- allows everyone to understand the purposes and roles of the production system, process, and operation
- enables you to find waste in the production systems, process, and operation

The use of function analysis to understand an equipment system:

- deepens your understanding of equipment

- allows everyone to understand the significance of equipment structure and motions
- enables you to find waste in methods of equipment design and use

The use of function analysis to understand an organizational system:

- deepens your understanding of the organizational system
- allows everyone to understand the structure and roles of the organization
- enables you to find waste in the structure and operation of organizational systems

The use of function analysis to understand an administrative and support system:

- deepens your understanding of the administrative or support system
- allows everyone to understand the structure and purposes of the administrative or support systems
- enables you to find waste in the structure and operation of the administrative or support systems

Using Function Analysis to Understand Systems

Inadequate understanding of systems gives rise to a variety of problems. Proceed according to the following steps when using function analysis to resolve these problems.

Step 1: Select an Object System

Select a target system about which understanding needs to be deepened and a consensus needs to be formed. Keep the following points in mind when you choose the system.

- Is understanding of the system inadequate?
- Is the system unclear?

- Is the system understood differently by different people, and from different perspectives?

Step 2: Form a Function Analysis Team

Next, put together a team of people who will perform function analysis of the object system. It may be appropriate for a single individual to carry out function analysis of a system involving only that one person's work. Most systems involve a number of people, however, and in such cases we recommend organizing the team around the people concerned and having them analyze the system. Team-based function analysis is extremely important to efficient analysis and shared understanding.

There should be about six to eight members on a team.

Step 3: Apply Function Analysis

Next, the individual or team analyzes system functions using function analysis worksheets.

Step 4: Use the Results of Function Analysis

Function analysis results in the completion of a function family tree. The process of constructing the function family tree corresponds to the process of understanding the system. The completed tree, moreover, shows the results of such an understanding.

Thus, when you want someone to understand the essence of a system, all you have to do is show him or her the system's function family tree. Moreover, when you want to use the results of function analysis to identify waste and problems in the system, all you have to do is clearly indicate types of waste and problems associated with each function on the function family tree.

Case Study: Using Function Analysis to Understand a System

A certain food company, Company P, applied function analysis to its corporate creed. Company P held the number-two spot in a particular area of the food field by virtue of integrated development, production, and merchandising. But corporate identity was unclear and the firm was complacent about its number-two position. At this point, it was decided that the 15-year-old corporate creed needed revamping, and so function analysis was applied to the current creed in order to restore its original significance.

Step 1: Select an Object System

Company P chose the corporate creed as the object of function analysis for the following reasons:

- There was no uniform, top-to-bottom understanding of the creed.
- As a result, principles for company actions were unclear.
- Company identity was weak.

The company decided to apply function analysis to its corporate creed both to reaffirm the significance of the creed and to spread its influence throughout the organization.

The corporate creed had been formulated as shown in Figure 7-2.

In Japan, we define corporate creed as the management precepts or doctrine of a company or association. Such creeds vary little from one company to another because every firm, no matter what the industry, needs to have its reason for existence. The creed presented in Figure 7-1, however, contains superficial language and provides little guidance for individual company actions and specific policies. It does not facilitate a systematic understanding of the creed itself.

Corporate Creed

We forever strive to achieve high quality, with reason and fairness, and with the ceaseless and unshakable determination to be winners. In this way, we discharge our corporate mission as a public institution and improve the welfare of all employees by promoting company prosperity.

1. *Quality is the life of Company P.*
   Trust begins when you build products infused with life and market them with sincerity.

2. *Communion of purpose is the essence of Company P.*
   Strength begins when you collect all views and debate every aspect of the objective.

3. *Discipline and inquiry are cornerstones of rapid progress.*
   Pleasure begins when you defy difficulties and confront tasks with creativity and ingenuity.

Figure 7-2. Company P's Corporate Creed

Step 2: Form a Function Analysis Team

As an expression of a company's precepts or doctrine, a corporate creed offers an ideal vision of the company to those both inside and outside the firm. This is why the entire organization — from top executives to individual supervisors — needed to participate in function analysis of the corporate creed. It was unrealistic, however, to expect that the thousands of people who made up the organization could all conduct function analysis and reaffirm the creed's significance.

Therefore, Company P put together a team composed of two people from top management (including the company president), two division managers, two department managers, and two supervisors. Each team member participating in the analysis was to represent others in his or her position but was not to use his or her position to impose views on others. For this reason, the leader of the function analysis team was chosen from among the department managers, on whose efforts the future of the company depended. A division manager was chosen as coleader.

Table 7-1 shows the composition of the team.

Table 7-1. Composition of Company P's Function Analysis Team

Leader:	• Department manager A (21 years experience in marketing, development, management planning)
Co-leader:	• Division manager A (29 years experience in development, manufacturing, personnel)
Members:	• President (38 years experience in marketing, development, accounting)
	• Director (34 years experience in development, manufacturing, management planning)
	• Division manager B (25 years experience in sales, accounting)
	• Department manager B (20 years experience in manufacturing, general affairs)
	• Supervisor A (12 years experience in development, manufacturing)
	• Supervisor B (11 years experience in marketing)

Step 3: Apply Function Analysis

The eight team members proceeded to reaffirm the corporate creed by analyzing its functions.

The application of function analysis ordinarily involves four steps. You begin by choosing a "theme," and then you understand the object of analysis and define and systematize functions.

In this instance, however, system design or system improvement was not the purpose of function analysis. The goal was to analyze the functions of the corporate creed "system," reaffirm it, and ensure that it was reflected in every company action and policy. This meant that there was no particular need for the theme selection and object of analysis steps used when designing or improving systems. In fact, the team began immediately with function definition. Figure 7-3 shows the results of this process.

Function Analysis Worksheet 4

Unit (Corporate Creed Phrase)	Functions	Criteria
We forever strive to achieve high quality, with reason and fairness	Strive to be winners. Have determination.	No criteria for degree of function achievement
With the ceaseless and unshakable determination to be winners	Pursue reason. Pursue fairness. Strive for high quality.	"
In this way, we discharge our corporate mission as a public institution and improve the welfare of all employees by promoting company prosperity.	Promote company prosperity. Improve employee welfare. Fulfill social mission.	"
Quality is the life of Company P. Trust begins when you build products infused with life and sell them with sincerity.	Strive for high quality. Build products infused with life. Market products with sincerity.	"
Communion of purpose is the essence of Company P. Strength begins when you collect all views and debate every aspect of the objective.	Build consensus. Consider all views. Debate all aspects.	"
Discipline and inquiry are cornerstones of rapid progress. Pleasure begins when you defy difficulties and approach work with creativity and ingenuity.	Improve oneself. Demonstrate creativity. Confront tasks.	"

Figure 7-3. Results of Defining Corporate Creed Functions

As worded, the corporate creed implied various functions to be performed. The team therefore divided the creed into phrases and proceeded to define the individual functions called for by each phrase.

Defining the functions of something like a corporate creed often reveals that words themselves serve to express functions. This doesn't mean, however, that the words used always accurately express the functions. For example, the following phrases were not workable expressions of function: "quality is the life of Company P," "communion of purpose is the essence of Company P," "discipline and inquiry are cornerstones of rapid progress." The team defined the functions involved by changing these phrases into true function expressions: "strive for quality," "build consensus," "improve oneself." In another example, the significance of the function of "infused with life" is lost unless you dig deeper into the meaning of the phrase.

When handling an object of analysis that, like a corporate creed, is neither a task nor a management system, the important thing is to pin down what is being said and to express it as a function.

Once the team members defined the functions, they proceeded to systematize them by goal/means relationships and to draw up a function family tree. Figure 7-4 shows the results of this process with Company P's corporate creed.

When systematizing goal/means relationships among the individual functions on the corporate creed's function family tree, the team needed to recheck the textual structure of the creed itself.

The "have determination" function was linked to the phrase "to be winners," so "strive to be winners" was its purpose. The "strive to be winners" function led to the "promoting company prosperity" phrase, so "promote company prosperity" was its goal. The "promote company prosperity" function was linked to the phrase "improve the welfare of all employees," so "fulfill social mission" and "raise employee welfare" became goal functions.

Pinning down goal/means relationships in this way identified four functions that employees needed to perform in order to reach the immediate corporate goal of "promoting company prosperity": (1) "strive to be winners," (2) "strive for high quality," (3) "build consensus," and (4) "improve oneself." These functions involve, respectively: (1) work attitude, (2) company trust, (3) organizational strength, and (4) job satisfaction.

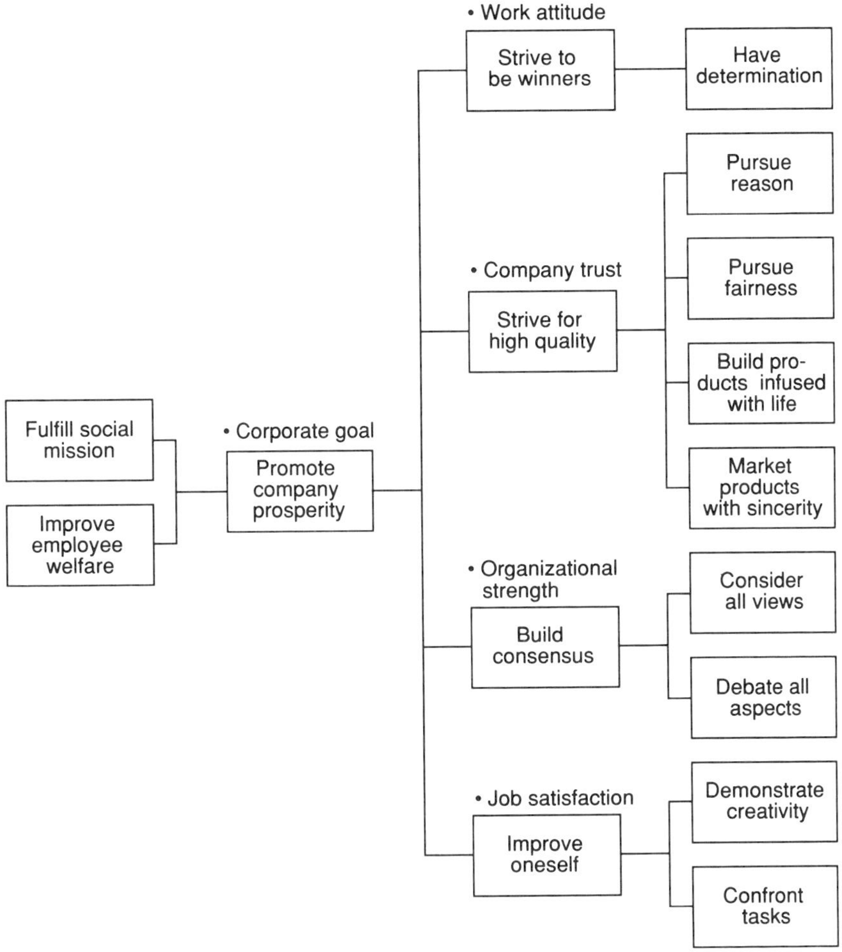

Figure 7-4. Function Family Tree of Company P's Corporate Creed

In addition, it became clear that the immediate corporate goal "promote company prosperity" served the two ultimate goals of fulfilling the company's social mission and improving employee welfare.

Step 4: Use the Results of Function Analysis

The use of function analysis resulted in the completion of a function family tree for Company P's corporate creed. The analytic process and the construction of the tree generated a systematic and shared understanding — whereas before there had been only a vague notion — of what the company stood for.

Put another way, a vague and superficial corporate creed was transformed through function analysis into corporate behavior that all could understand. At the same time, the process brought company identity into sharper focus.

The clarification of the corporate creed revealed the linkage between company policies and the ideal vision of the company expressed in the creed.

Company P had, for example, introduced TQC methods some time earlier and was in the process of expanding QC circle activities. But there was no companywide shared understanding of the significance or goals of QC circles. Plotting QC circle goals on the corporate creed's function family tree guided everyone to a clear understanding of the purpose of QC circles. In other words, this process gave the growth of QC circles an explicit basis in the corporate creed. The goals of QC circles were now clear: (1) to pursue reason and thereby enhance trust in the company through striving for high quality; (2) to consider all views and debate all aspects of problems, thereby enhancing the organizational strength of the company through consensus building; and (3) to demonstrate creativity and confront challenging tasks, thereby enhancing work satisfaction through self-improvement.

FUNCTION ANALYSIS AND SYSTEM OBJECTIFICATION

Why You Need to Objectify Systems

Recall the distinction we made between two kinds of systems: hardware systems, such as products or equipment, and software systems, such as organizations or administrative tasks.

Such systems are consciously designed by individuals or organizations and therefore always have some purposes, goals or functions. It is not unusual, however, for the goals or functions of the system to be unclear to people involved with the system. Why? Because there is no accurate method for representing system goals and functions objectively.

Consider the case of a product system. The object itself or drawings of the object show only the system's physical aspects. In a support system, too, documents and flowcharts for tasks in the system show only the flow or methods of work.

Figure 7-5 lists various diagrams and methods used to represent different types of systems, but not one of these does more than show apparent — or phenomenological — aspects of the system in question. The result is that it is impossible to gain an accurate understanding of system goals and functions.

System	Traditional method for objective display
• Product system.............	Actual object and drawings (method shows only physical aspects of system)
• Production system.........	Process analysis charts (method shows only observable aspects of system)
• Equipment system.........	Equipment diagrams (method shows only physical aspects of system)
• Organizational system...	Organizational charts (method shows only structure of system)
• Sales system...............	Task flowcharts (method shows only flow of system)

Figure 7-5. Traditional Methods for Objective Display of Systems

Applying function analysis to systems results in the construction of function family trees. These trees give high visibility and objective expression to functions the system needs to perform.

To put it another way, analyzing systems through function analysis leads to the construction of function family trees, which show functions — the essence of systems — in an objective manner.

System Objectification and the Effects of Function Analysis

Function analysis is an approach to systems from the point of view of their functions. In other words, it (1) is an analysis of system goals and functions and (2) results in an objective representation of that system's essence.

The use of function analysis to objectify a system promotes understanding of the system and improves interdepartmental communication.

The use of function analysis to objectify a product system:

- allows everyone to understand the product's design concept
- improves communication among product-related departments by developing a common understanding of the design concept via a function family tree

The use of function analysis to objectify a production system:

- allows everyone to understand the essence of the production system
- improves communication among production-related departments by allowing everyone to share an understanding of the essence of the production system via a function family tree

The use of function analysis to objectify an equipment system:

- allows everyone to understand the essential significance of equipment
- improves communication among equipment-related departments by allowing everyone to share an under-

standing of the essence of equipment via a function family tree

The use of function analysis to objectify an organizational system:

- allows everyone to understand the roles of the organization and the people who constitute it
- clarifies who is supposed to do what
- improves communication among departments by promoting mutual understanding of the essence of the organization via a function family tree

The use of function analysis to objectify an administrative and support system:

- allows everyone to understand goals representing the essence of administrative and support systems
- improves communication among administrative and support-related departments by promoting mutual understanding of the essence of support activities via a function family tree

Case Study: Using Function Analysis to Objectify a System

The use of function analysis to objectify a system may mean objectifying either an existing system or a new system. In either case, the process involves analyzing the functions of the system itself and then representing those functions on a function family tree. The procedure here is the same for system understanding. As an example, let's consider the application of function analysis to a sales system.

Changing market demands and the accelerating pace of innovation placed a severe competitive burden on Company F, a manufacturer and marketer of office machines. Competition on the sales front had stiffened as well, the sales division being engaged in a fierce marketing war. The company was beefing up its sales force, but that strategy had yet to produce increased sales.

An overhaul of the sales system was clearly needed. To that end, the company decided to analyze the functions the sales system needed to carry out and to use a function family tree to determine objectively what form its sales system should take. Although a sales system was already in place, Company F applied function analysis as though it were building a system from scratch.

Step 1: Select an Object System

The sales system was chosen as the object of function analysis for the following reasons:

- The sales force has been increased, but sales were not rising proportionally.
- It was not clear what constituted sales activities.
- The sales organization was not growing.
- Sales efficiency was low.

The decision was made to apply function analysis to the sales system as a prerequisite to a complete overhaul of the system. A function family tree would show what the sales system should look like.

Step 2: Form a Function Analysis Team

Since a team had already been put together to strengthen the sales force, the same team was chosen to carry out function analysis. The company president, who was also head of the sales division, served as leader; the director in charge of sales acted as coleader; and seven department heads in the sales division served as the other team members. These seven carried out the function analysis.

Step 3: Apply Function Analysis

The seven department heads decided to analyze the functions of the sales system and then use a function family tree to determine what the sales system should look like.

Following function analysis procedures, they began by understanding the object of analysis. Figure 7-6 shows the function analysis worksheet the team used at this stage of understanding.

As the requirements list shows, the goal of overhauling the sales system meant that the company had to secure a 30-percent market share by meeting periodic sales targets.

To meet those targets, $1 million was allocated for the initial phase of investment in reforming the sales system. In other words, $1 million was the allowed cost of investment for changing to a new sales system. At the same time, the new system had to be established within the framework of current organizational structures and current staffing levels.

The next step was to fill out an activities design description on the basis of the requirements list. Note that detailed requirements — and quantitative breakdowns — have been omitted from the requirements list in Figure 7-6. Figure 7-7 shows the support activity design theme description.

In this example, items on the theme description are exactly the same as on the requirements list (see Figure 7-6), but this is because the two worksheets have been coordinated. If the theme description has not yet been coordinated with the requirements list, the two must be matched (that is, a tradeoff must occur) before the theme description can be completed. Once again, details have been omitted on the theme description sheet.

A clear support activity design theme description marked a full understanding of the object of analysis. (See Figure 7-8.)

At this point, the team moved on to the next step: function definition.

Function Analysis Worksheet 1

Requirements List for Sales System	
System Requirements	
• Goals and objectives of work............	Maintain 30% market share in office equipment field
• Departmental requirements..............	Reliable communication of customer demands (development)
	Swift and reliable identification and communication of customer claims and complaints (quality assurance, manufacturing)
	Reduced delivery schedule changes (production control)
• Goals and objectives of system.......	Meet target sales increase each period
System Design Requirements	
• Rules and regulations.......................	None in particular
• Allowable investment........................	Up to $1 million for communications equipment
• Other design constraints...................	Organizational structure will not change
System Operating Requirements	
• Volume of work and required time....	Same as now
• Quality and precision of work............	Level orders; no more than 5% should fluctuate
• Personnel	Current staff
• Other operating constraints...............	No product discounts, bills approved

Figure 7-6. Requirements List for Sales System

Function definition involved identifying and evaluating activities to achieve task goals and then defining suitable activities as functions. The result of this process is shown in Figure 7-9.

Using brainstorming techniques to circumvent preconceived notions, the team members freely generated ideas for

Function Analysis Worksheet 2

System Design Theme Description

1. System name
   Sales system

2. System requirements
   (a) Work goals and objectives
   Meet target sales increase each period
   (b) Departmental requirements
   - Development: Reliable communication of customer demands
   - Quality assurance: Swift and reliable identification and communication of customer claims and complaints
   - Manufacturing: Swift and reliable identification and communication of customer claims and complaints
   - Production control: Reduced delivery schedule changes

3. System design requirements
   (a) Rules and regulations
   None in particular
   (b) Allowable investment
   Keep outlays below $1 million for communications equipment, etc.
   (c) Other
   Current organizational structure and personnel will not change

4. System operating requirements
   (a) Volume of work and required time
   Same as now
   (b) Quality and precision of work
   No more than 5% of orders should fluctuate; level orders to within 20%
   (c) Personnel
   Operate with current personnel
   (d) Other
   Offer no product discounts; approve no bills

Figure 7-7. System Design Theme Description

activities that would help them achieve task goals and objectives. (See Figure 7-8.) At this stage, they needed to consider methods from as broad a perspective as possible.

Next, the team evaluated each activity proposed to determine whether or not it satisfied each of the requirements (constraints). For example, item 10, "liberalize payment terms,"

Function Analysis Worksheet 3

Support system name: Sales system
Work goals and objectives: Meet target sales increase each period

Support Activities to Achieve Goals and Objectives	Evaluation
1. Develop new customers	○
2. Advertise the product using all available media	○
3. Grasp the essence of customer complaints and respond swiftly	○
4. Reflect changes originating in customer complaints in next period's product	○
5. Conduct thorough after-sales service	○
6. Increase replacement rate	○
7. Observe delivery schedules strictly	○
8. Provide support so that product reliability can be preserved	○
9. Be aggressive in providing new products	○
10. Liberalize payment terms	×
11. Understand customer needs (i.e., customer demands) rapidly	○
12. Use opinion leaders	○
13. Increase sales force	×
14. Increase affiliates	○
15. Increase sales outlets	×
16. Increase discount rates	×
17. Enhance incentive system	×
18. Increase samples	○
19. Sell with accessories	○
20. Have top officers visit customers	×
21. Hold exhibitions	○
22. Increase visits to customers	○
23. Arrange for customers to visit company	○
24. Keep up with trends at competitor companies	○
25. Use own products in-house	○
26. Issue exhaustive interim reports	○

Figure 7-8. Support Activities to Achieve Goals and Objectives Evaluation

did not satisfy the requirements of running the system and so was rejected (and marked with a "–"). Item 13 ("increase sales force") was rejected for the same reason. In this way, the team pared down the number of activity ideas it had generated by roughly 30 percent.

Function Analysis Worksheet 4

Support Activity	Functions		
	Verb	Noun	Criteria
1. Develop new customers	Increase	New customers	Omitted
2. Advertise product using all available media	Inform of Improve	Products Product image	"
3. Grasp essence of customer complaints and respond swiftly	Understand Communicate	Complaints Complaints	"
4. Reflect changes originating in customer complaints in next period's product	Understand Communicate Resolve	Complaints Complaints Complaints	"
5. Conduct thorough after-sales service	Support Understand Predict Resolve Prevent	Product use Complaints Complaints Complaints Complaints	"
6. Increase replacement rate	Continue	Merchandise transactions	"
7. Adhere strictly to delivery schedules	Adhere to	Delivery schedules	"
8. Provide support so that product reliability can be preserved	Inform of Instruct in	Product use Product use	"
9. Be aggressive in providing new products	Announce Broaden	New products Choice of products	"
10. Understand customer needs (i.e., customer demands) rapidly	Understand Sell Communicate	Customer demands Products Customer demands	"

(etc.)

Figure 7-9. Defining Support Activities as Functions

Next, the team proceeded to define support activity functions by checking the sense of each activity and then expressing it as a function, with a verb-noun combination. (See Figure 7-9.)

The first activity — that of developing new customers — meant performing the function of "increase new customers."

The second activity ("advertise product using available media") meant performing the function of "inform of products" in relation to customers. This particular activity, however, encompassed more than simply informing. It meant satisfying the function of improving product image in order for the product to sell.

The third activity ("grasp essence of customer complaints and respond swiftly") meant performing two functions: understanding and communicating customer complaints. The term *swiftly* was a constraint on both understanding and communicating and therefore needed to be included in the function expressions.

The fourth activity ("reflect changes originating in customer complaints in next period's product") referred to the achievement of nearly the same functions as the third activity. The fifth activity ("conduct thorough after-sales service") involved performing a general function designated "support product use." It was also necessary, at the product-use stage, to satisfy other functions, such as "understand complaints" and "resolve complaints."

As we noted in Chapter 6, it is crucial to convert activity descriptions into function expressions that are universally understandable. Also important is getting at the root purposes and roles of tasks through adequate verification and a nonrestrictive stance on work methods.

We will omit discussion of the establishment of function criteria for this example.

Once the team had a clear grasp of support activities and their functions aimed at achieving periodic sales targets, it moved on to the next step: systematizing the goal/means relationships among functions. Figure 7-10 shows the standard result of such systematization — a function family tree. Let's look at the approach the team took in systematizing the goals and means for individual activity functions.

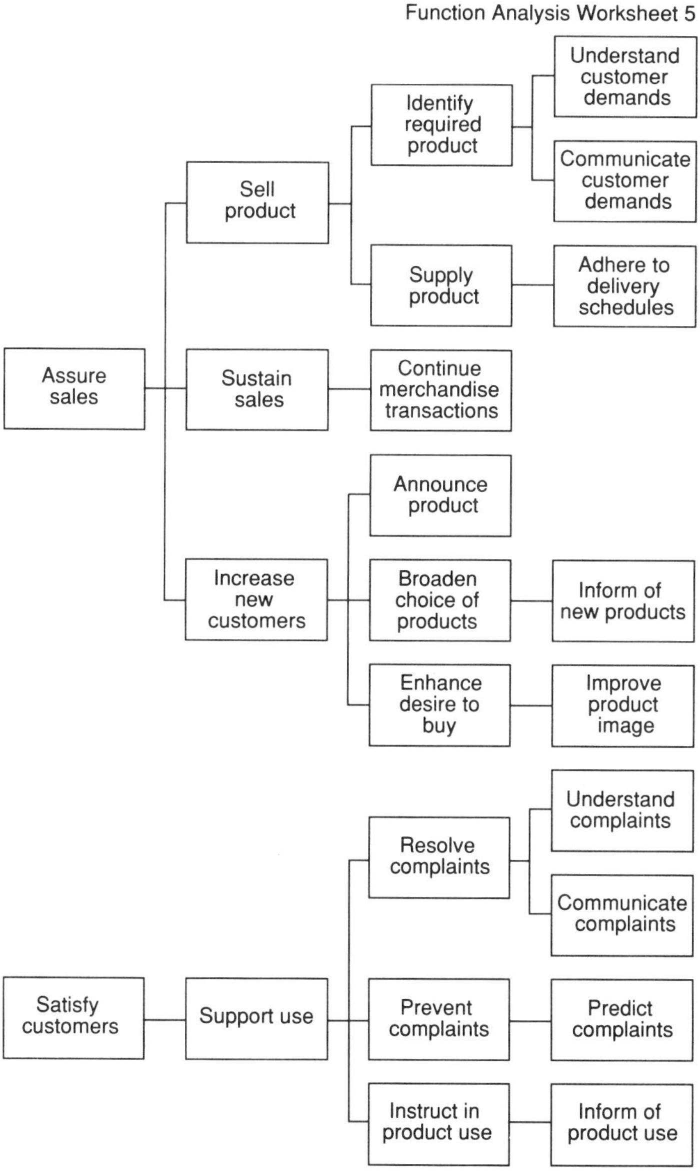

Figure 7-10. Function Family Tree for Support System

First, the goal of the two functions designated "understand customer demands" and "communicate customer demands" is determining what the customer wants. You can express this as the function "identify required product." This function, in turn, serves the goal of selling the customer what he or she needs ("sell product"), a function whose goal is "assure sales."

We have stated that the "identify required product" function serves the function of selling products. However, that identification also needs to be communicated accurately to the development department if the required product does not already exist. Identification of the required product, in other words, serves an additional function, "develop new products," in order to meet the development department's requirement for accurate communication of customer demands. This departmental function, however, has been left off the function family tree.

The "adhere to delivery schedule" function serves the function of supplying the required product ("supply product"), which in turn serves the function of "sell product."

By identifying each function's goal and systematizing the goal/means relationships among functions, the tree reveals objectively that three functions are needed to perform the ultimate function of "assure sales": the direct "sell product" function and then the functions designated "sustain sales" and "increase new customers."

These are functions required of a sales system from the perspective of company management. A sales system must also perform external functions with respect to other players, such as the market and customers.

When complaints come in from customers, the functions designated "understand complaints" and "communicate complaints" are needed to satisfy the "resolve complaints" function. This last function is needed in relation to customers, in turn, to perform the "support use" function. And supporting customer use of the product is ultimately for the purpose of satisfying the customer ("satisfy customer").

"Support use" and its goal, "satisfy customer," are both functions a sales system needs to perform in relation to the market and customers outside the company.

This systematization of goal/means relationships showed objectively that the two ultimate functions a sales system must perform are those of assuring sales and satisfying customers. It also clearly identified what intermediate ("means") functions needed to be satisfied in order to perform those ultimate functions. Although we have omitted such details from our example, adding relevant numerical criteria above each function on the function family tree results in a chart that shows what the sales system should look like (that is, what functions need to be performed to what extent). The representation is objective, visual, and easy to understand.

Constructing a function family tree thus identifies goals that constitute the essence of the sales system. It makes those goals understandable to everyone and improves communication among sales-related departments by facilitating mutual and shared understanding.

Step 4: Use the Results of Function Analysis

For Company F, function analysis resulted in a completed function family tree and in an objective picture of how the sales system should look.

In contrast to the extremely abstract first-order tree functions (that is, the ultimate functions of the sales system) expressed as "assure sales" and "satisfy customers," fourth-order functions ("understand customer demands," "communicate customer demands," and so on) designate specific means.

In our example, Company F used this sales system function family tree to gain an objective understanding of what the system should look like. It went on to use the tree as a basis for overhauling the sales system by considering new work methods for performing each of the functions shown.

After examining new work methods, the company moved on to design. First, it established methods for performing each of the functions shown at the extreme left of the tree. We will omit discussion of the new methods considered. Suffice it to point out, however, that in overhauling the sales system, Company F was able to use the function family tree to show objectively how the system should look. This allowed everyone to understand the tasks to be designed and thereby facilitated effective system design.

FUNCTION ANALYSIS AND SYSTEM IMPROVEMENT

What Is System Improvement?

System improvement means transforming a current system into a better one. A system may be a "hard" system, such as a product or equipment system, or it may be a "soft" system, such as a production, support, or service system.

There are also various kinds of system improvement. Aspects of a system that may be improved include (1) technical components, such as functions or quality, (2) economic components, such as costs, and (3) time components, such as delivery schedules and capacity. Improvement might mean increasing quality or functions, lowering costs, or shortening delivery periods. It may also include enhancing factors such as safety. (See Figure 7-11.)

We see, then, that the term *system improvement* refers to actions that include enhancing system function and quality, reducing costs, shortening delivery periods, and enhancing safety.

No system in existence is perfectly problem-free. All systems suffer from problems to one degree or another, and these problems involve the kinds of issues just mentioned. System improvement consists of replacing the current system with a new system in which such problems have been resolved.

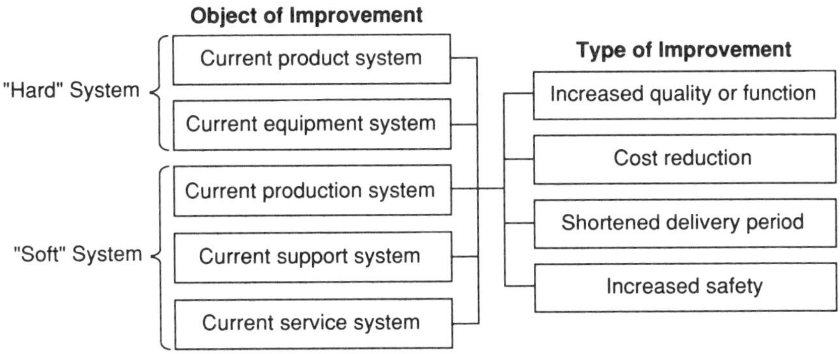

Figure 7-11. What Is System Improvement?

Using Function Analysis to Improve Systems

Using function analysis to improve a system involves, first of all, analyzing the functions of the system and constructing a function family tree. The tree is a systematic diagram showing what functions the target system needs to perform. From the functions represented on the function family tree, you select a function needing improvement, generate ideas for improving that function, and draw up improvement plans. Figure 7-12 traces the steps for constructing improvement plans based on the results of function analysis.

Case Study: Using Function Analysis to Improve a System

System improvement takes a variety of shapes. The system that is the object of improvement, for example, may be a product or equipment "hardware" system, or a production or support "software" system. The needed improvement, moreover, may involve issues such as function or quality enhancement, cost cutting, delivery period reduction or safety enhancement.

A certain foodstuffs company used function analysis to improve its production system (that is, the manufacturing

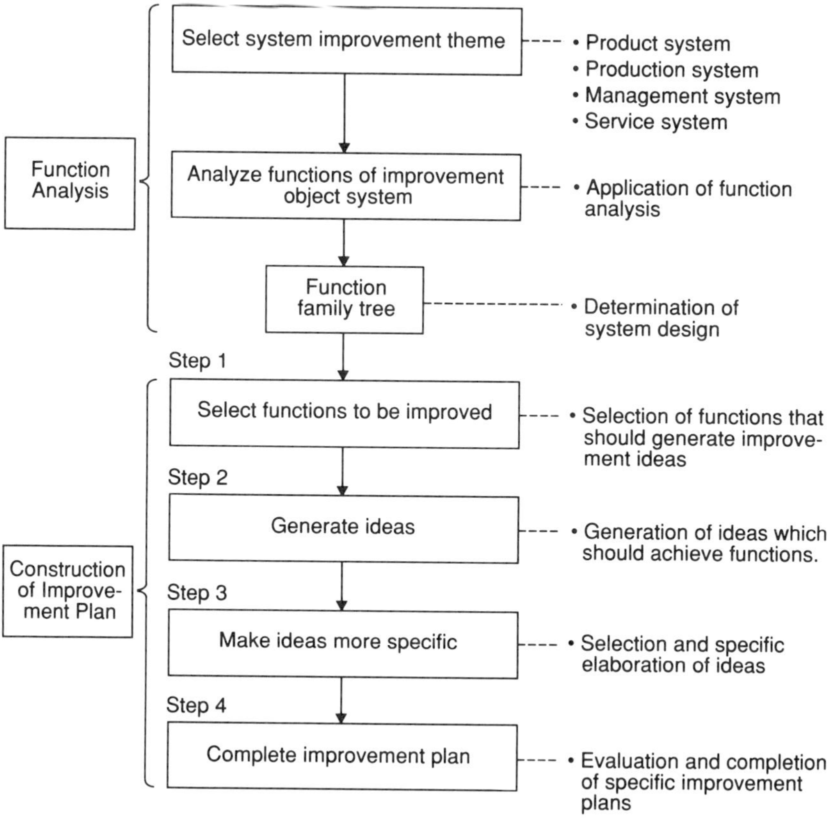

Figure 7-12. Procedure for Using Function Analysis for System Improvement

process). As an improvement theme, the company chose a sliced ham manufacturing process in which it wanted to see higher yields (that is, improved quality) and lower costs.

Step 1: Select a Process Improvement Theme

The sliced ham process chosen as a theme consisted of a string of subprocesses in which frozen ham was treated, sliced by machine, and then packaged.

The process was selected as a target of improvement because of poor yield from the raw material which, given the high production volume, added up to millions of dollars lost each year. The objective in applying function analysis was to cut yield losses in half. As a first step in using function analysis to improve the process, the company drew up the process description shown in Figure 7-13.

Function Analysis Worksheet 1

Process Improvement Theme Description
1. Summary and scope of process improvement theme • Process: from treating material to secondary packaging of sliced ham • Scope of improvement target: thawing, treating, salting, molding, heating, cooling, secondary packaging 2. Reasons for selecting process improvement theme • Improvement will have a substantial effect because of rising production volume at Plant T. • Improvement will have a substantial effect because both rate and value of losses are high. • The process is a major cost reduction target at Plant T. 3. Improvement targets • 50% reduction in rate and value of losses • *X*% reduction in total plant costs

Figure 7-13. Process Improvement Theme Description

Step 2: Understand the Object of Analysis

Accurate function analysis requires a correct understanding of the object of analysis, so the company collected various kinds of information pertaining to the process to be analyzed.

The information related to factors such as specifications and quality of the final packaged product, the sliced ham process itself, individual operations in each of the subprocesses, and cost and productivity yields.

Understanding information on product specifications and quality was a prerequisite to improving the process. Information on processes and operations facilitated a clear grasp of the status quo and provided clues to the functions of those processes and operations. Information on yields enabled team members to develop a detailed understanding of types of yield losses and served as data for identifying processes and operations that were generating problems.

The company gained an accurate understanding of the object of analysis by collecting and organizing the information described above in accordance with an information collection plan. Figure 7-14 shows a simplified list of the categories of information collected.

Function Analysis Worksheet 2

Information to Be Collected

1. Product information
   - Product specifications for sliced ham (governmental and company specifications)
   - Product quality standards for sliced ham and packaging (company standards)
   - Details of market claims relating to sliced ham (data for previous year)

2. Process information
   - Detailed sliced ham process analysis sheet
   - Sliced ham process layout diagram
   - Chart of specified equipment, tools, and dies for sliced ham process
   - Process capacity charts for sliced ham process

3. Operational information
   - Work standard charts for each operation
   - Individual layout diagrams for each operation

4. Productivity and cost information
   - Standard and actual times for sliced ham process
   - Actual process yield for sliced ham process (previous 12 months)
   - Rate and value of yield losses for sliced ham process (previous 12 months)

Figure 7-14. Information to Be Collected

Step 3: Define Functions

The objective in improving the sliced ham process was a 50 percent reduction in yield losses. This target would have been difficult to achieve by merely generating ideas and constructing improvement plans based on functions identified at the level of individual tasks or operations. Indeed, such an approach would have resulted in only isolated improvements. Therefore it was decided to identify process-level functions and use them as the basis for improvement plans. This would make it possible to carry out radical improvements in the process and to attain the project target. Figure 7-15 shows the results of defining the functions of constituent components at the level of individual processes.

A manufacturing process performs the function of transforming some object. In these terms, the function of transforming frozen meat into thawed meat can be designated "melt ice." Transforming the thawed meat into refined, well-formed units required the performance of several functions, such as "remove fat," "remove bones," and "remove foreign matter." The "well-formed" stipulation also meant carrying out the function defined as "shape meat."

The molding process involved transforming flavored meat into cylindrical loaves and essentially had to perform the function designated "form meat." It also had to fulfill partial functions such as "fix meat thickness" and "fix meat length." Making meat of various lengths uniform required a "join loaves" function, and a "remove air bubbles" function was needed to keep air bubbles from the molding process out of the meat.

This was how the definition of needed functions proceeded, with attention being paid to the state of the product before and after each process.

Our example does not show all constraints on the functions indicated, but important criteria for the treating process were

Function Analysis Worksheet 4

Process and *What* chart		Function		Criteria
		Verb	Noun	
Frozen meat				
	Thawing process	Melt	Ice	Internal temperature of 5-6° C
Thawed meat				
	Treating process	Remove Remove Remove Shape	Fat Bones Foreign matter Meat	Fat layer of $\leqq$ 3 mm 0 bones 0 foreign matter
Refined, well-formed meat				
	Salting process	Inject Add Improve	Brine Flavorings Meat color	
Salted meat				
	Massaging process	Spread out Tenderize Eliminate Improve Retain	Flavoring Meat Air bubbles Meat color Brine	No air bubbles
Flavor-permeated meat				
	Molding process	Form Fix Fix Join Eliminate	Meat Meat thickness Meat length Loaves Air bubbles	Circular... 30.5 ± 0.2 cm in circumference 1.3-1.6 m in length Vacuum...
Well-formed meat				

(etc.)

Figure 7-15. Process and *What* Chart

crucial in determining product quality. "Remove fat," for example, is accompanied by a criterion indicating how much fat is to be trimmed: "fat layer of ≤3 mm." The "remove bones" and "remove foreign matter" functions need to be accomplished to the extent indicated: "0 bones" and "0 foreign matter." Such criteria are quality standards for the product and process.

Once the needed functions of each of the processes in the scope of the improvement theme was identified, the company moved on to the next step: function systematization.

Step 4: Systematize Functions

Figure 7-16 shows a completed function family tree for the sliced ham process. The tree was constructed as a result of systematizing all needed functions by their goal/means relationships.

Function Analysis Worksheet 5

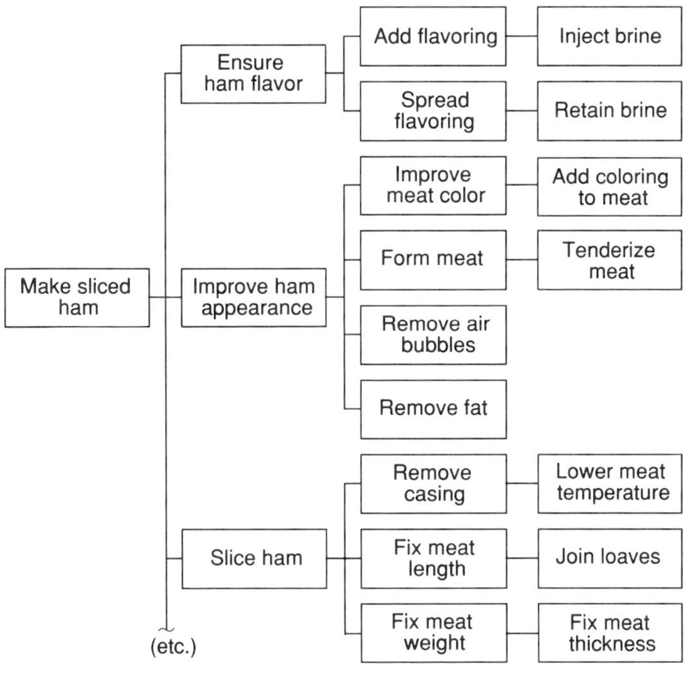

Figure 7-16. Function Family Tree for Sliced Ham Process

Looking at goal/means relationships on the function family tree, we see — apart from the first-order function designated "make sliced ham" — three functions the process must perform at the secondary level: "ensure ham flavor," "improve ham appearance," and "slice ham." The company, then, understood that these were the basic functions the sliced ham process had to fulfill. Omitted from the function family tree shown here are other second-order functions noted on the original tree: "ensure meat quality," "preserve meat freshness," "guarantee manufacture date," and so forth.

Systematizing functions yielded a clear picture of what the sliced ham process should look like and completed the preliminary requirements for process improvement. Note that, in the example, the criteria recorded above each function on the function family tree have been omitted.

The next step was to draw up improvement plans based on the function family tree.

In the first step of plan construction — choosing functions to be improved — the company broke down yield losses (in dollars) by second-order function. This exercise revealed that most losses were being generated by the process performing the "slice ham" function. Significant losses were also identified at processes carrying out the functions designated "improve ham appearance" and "assure meat quality." The company concluded that problems existed in the performance methods — in the processing methods, equipment, and so forth — used to achieve function goals. It therefore generated ideas for new ways to perform each function: new processing methods and equipment that would keep losses from occurring.

The next step in plan construction was to generate ideas for improving each of the three target subprocesses: "slice ham," "improve ham appearance," and "assure meat quality." This exercise resulted in numerous goal-oriented ideas for radical process changes.

At step 3 in the construction of improvement plans, the company identified the most valuable proposals by evaluating all goal-oriented ideas generated in step 2 in terms of their effectiveness and technical feasibility. The most promising ideas were then incorporated into specific improvement plans. Step 4 of plan construction involved organizing those plans by process and implementing them.

To summarize the process, using function analysis enabled the company to make previously impossible function-based process improvements and to exceed its goal of loss reduction.

FUNCTION ANALYSIS AND SYSTEM DESIGN

What Is System Design?

System design is an activity that begins with a recognition of the goals or purposes of the product or system to be designed. One then creates methods to satisfy those goals or purposes, evaluates the methods in terms of factors such as quality, cost, and delivery schedules, and finally displays them in a form accessible to everyone. The object of design may be a product, equipment, or other "hardware"; or it may be "software," such as production, support, or service systems.

System design is a process that (1) transforms customer demands into specific drawings and methods, (2) assigns system goals to objects, information, and activities, (3) processes information, and (4) involved decision-making. (See Figure 7-17.)

Finally, system design produces new products and services that satisfy customer demands in terms of system goals, quality, cost and delivery schedules, and so forth.

The foregoing definition of system design illustrates the crucial importance of correctly translating customer demands into system goals and recognizing those goals. Function analysis is a method aimed at that recognition. In other words, function analysis is a technique serving as the starting point for system design.

Moreover, using function analysis to gain an accurate understanding of system goals allows you to consider creative, goal-based (that is, function-based) methods and clarifies criteria for evaluating designed systems.

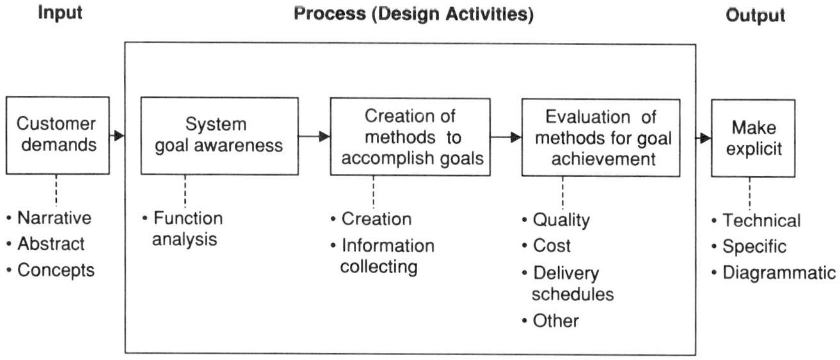

Figure 7-17. What Is System Design?

For these reasons, function analysis can be considered a basic technique in system design.

Using Function Analysis to Design Systems

Correctly understanding the object of design — especially system requirements — is the first step in using function analysis for system design. Those requirements form the basis for constructing a function family tree, a diagram systematically showing all functions the object of design needs to perform. Methods for carrying out functions included on the function family tree are generated as ideas, which are then evaluated and incorporated into the system.

Figure 7-18 shows these procedures in schematic form.

Case Studies: Using Function Analysis to Design Systems

As we have already mentioned, system design may focus on a variety of systems, from product or equipment "hardware" systems to "software" systems such as processes, operations, support activities, or organizations.

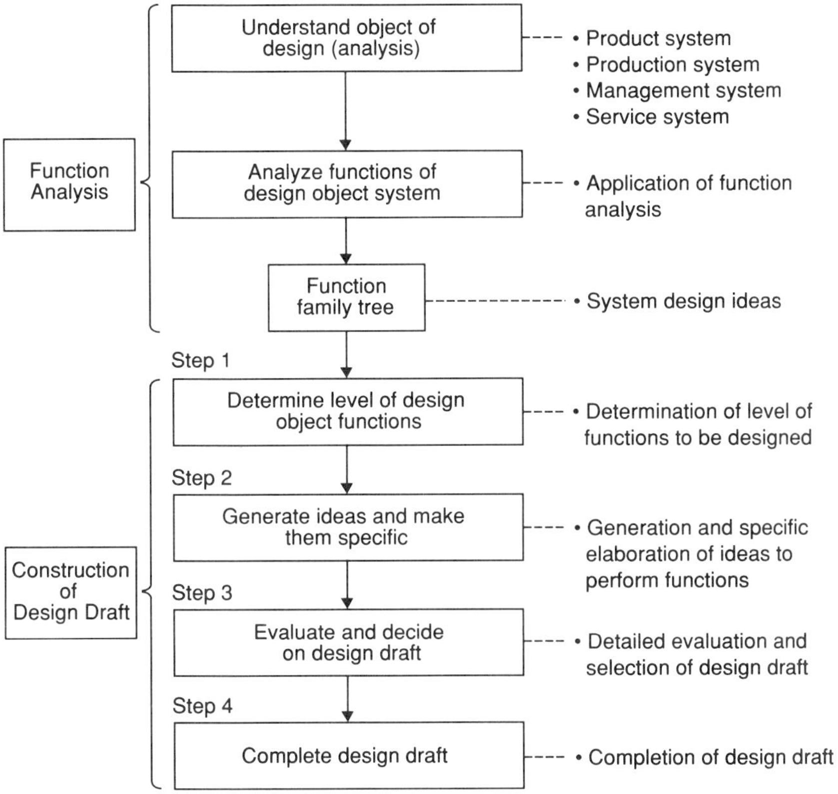

Figure 7-18. Procedure for Using Function Analysis for System Design

To more closely examine the application of function analysis to system design, we will look at a product ("hardware") system and a support activities ("software") system.

Function Analysis Applied to the Design of a Steam Iron (Hardware)

Company S is a major manufacturer of household appliances caught up in the competition to develop new appliances, a mature market. Division T of Company S produces small household appliances, but in the dizzying race to develop new products was not performing as well as hoped. The division was struggling to increase its share of the household iron market. Since high-end steam irons were the mainstay of the household iron line, the company chose to use function analysis to develop a new advanced steam iron.

Step 1: Understand the object of analysis. A specifications sheet had already been established for a high-end steam iron, so the company began by checking over those specifications and confirming information wherever necessary. The requirements list shown in Figure 7-19 is the result of this preliminary work.

On the basis of the requirements list, the company compiled a description of the product design theme. Information in the description follows the requirements list fairly closely, so we have omitted the details in our example. In the case of a design theme such as that for the steam iron, where a product is aimed at an indeterminately wide range of users, it is important in terms of understanding requirements to collect and organize exhaustive information on the various levels of product use.

Once the requirements list and theme description were clear, the company moved on to function definition.

Step 2: Define the function. In the function definition stage, each function needed for the target product was defined based on requirements identified in the previous stage. Requirements selected for function definition were chosen on the basis of their relevance to product function from those on the requirements list and the theme description.

Figures 7-20 and 7-21 show how steam iron functions were defined based on identified requirements.

Function analysis generally proceeds in three stages: (1) selecting requirements for function definition, (2) defining the functions of those requirements, and (3) establishing function criteria. This procedure was followed in the steam iron example.

Function Analysis Worksheet 1

Requirements List

Use-related Requirements

Users	Single or married women 20 to 50 years of age
Use environment	On *tatami* or tables
Performance	• Glides over fabric easily; allows ironing in tight spaces (e.g., around buttons) and sheets, etc. • Weight is suitable (1-1.5 kg); allows steaming over a wide area • Base plate temperature is uniform and adjustable over the range of 80°-210° C • Iron reaches set temperature within 1 minute 50 seconds • Can steam constantly for 15 minutes (water tank holds 180-200 cc) • Iron can be switched back and forth from steam to dry settings • Water level is easy to read • Uses 100 V AC, 50/60 Hz power source
Reliability	• Product life is 1,000 hours or more (30 minutes/day × 5 years) • Water does not leak out when iron is placed upright during use • Water does not leak out when temperature falls
Maintainability	• Can be stored after use as is without rusting • Water residues can be cleaned out • Iron is easy to fill
Operability	• Has "Power" indicator, "Ready" indicator • Cord does not get in the way during use • Cord does not get in the way during storage • Can be stored immediately after use; is easy to store
Safety	• Iron cannot cause burns or fires by being tipped over • Iron cannot cause burns or fires by being left on • Iron is designed not to harm children who may knock it over or bump into it

Sales-related Requirements

Differentiation, selling points	• Color and shape are appealing to users • Design makes ironing enjoyable • Integrated steamer
Distribution/ shipping	• Conforms to existing irons on the market
Price	• $164
Date & no. available	• April 1989; 100,000 units planned

Design Requirements

Regulations, etc.	• Design conforms to legal appliance standards
Cost	• Target cost of $XX

Figure 7-19. Requirements List for a New Steam Iron

Function Analysis Worksheet 3

Requirements	Description
Use goals	• The iron is designed to be used by individuals 20 to 60 years of age for ironing clothing and sheets, etc., on *tatami*, or tables. • The iron must be easy to use, easy to operate, safe, and easy to store.
Use environment	• Used on *tatami* or tables
Performance	• Base temperature is uniform (± 5%) in the range of 80°-210° C • Glides easily over fabric; facilitates ironing in tight spaces • Weight is suitable (1-1.5 kg); steams over entire plate • Reaches needed temperature within 1 minute 50 seconds • Can steam constantly for 15 minutes (water tank holds 180-200 cc) • Water level in tank is easy to read • Can easily be switched back and forth from steam to dry settings • Uses 100 V AC, 50/60 Hz power source

(etc.)

Figure 7-20. Selecting Requirements for a New Steam Iron

The selection of requirements involved identifying various requirements as needed product functions. These requirements included demands on use goals, use environment, performance, reliability, maintainability, operability, safety, differentiation, regulations, and standardization. In our example, we illustrate some of the requirements identified in the areas of use goals, use environment, and performance.

Defining requirement functions involved treating use goals and use environment as the same category of requirement and then defining functions. Here, the ultimate goal of the iron is "straightening out clothing" — including garments, sheets, and so forth. In terms of use environment, the requirement that the irons be usable on either *tatami* mats or tables means satisfying the function designated "maintain iron stability."

Function Analysis Worksheet 4

Requirements	Functions		
	Verb	Noun	Criteria
(Use goals) • The iron is designed to be used by single or married women 20 to 60 years of age for ironing clothing and sheets, etc., on *tatami* or tables • The iron must be easy to use, easy to operate, safe, and easy to store	Straighten out Maintain Facilitate Maintain Facilitate	Clothing Iron stability Ironing Safety Storage	• Fabric types... • Stability on *tatami* and tables • Cord does not get in way
(Performance) • Base temperature is uniform • Glides easily over fabric; facilitates ironing in tight spaces • Weight is suitable; steams over entire plate • Reaches needed temperature rapidly • Can steam continuously • Can easily be switched back and forth from steam to drysettings • Uses household current as power source	Maintain Make Allow easy sliding of Facilitate Exert Generate Hasten Generate Switch between Supply	Needed temperature Base plate temperature uniform Base plate Close-quarters ironing Pressure Steam Heating Steam Steam and dry Electrical power	• 80°-210° C • ±5° variability • Around buttons • 1-1.5 kg • Over entire plate • In 1 minute 50 secs. • 15 minutes or more • x number of times • 100 V AC, 50/60 Hz

(etc.)

Figure 7-21. Defining Functions of those Requirements

In regard to performance requirements, the demand that the base plate temperature be uniform means satisfying the function designated "maintain needed temperature" to "straighten out clothing." It also means satisfaction of the "make base plate temperature uniform" function.

Requirements that the iron glide smoothly on fabric and facilitate ironing in tight places depend on characteristics of the base plate. They mean satisfying the functions designated "allow easy sliding of base plate" and, of course, "facilitate close-quarters ironing." Requirements that the iron be of a suitable weight and that steaming be possible over the entire base plate mean satisfying the functions designated "exert pressure" and "generate steam."

By checking the meaning of each requirement in this way, the company defined functions the product would have to perform.

Next, Company S established criteria by associating various conditions and degrees of function achievement with individual functions on the requirements list. As the example shows, clear quantitative criteria can be established for the functions of a tangible product. These quantitative function constraints are established at the level of product use based on user demands.

After using requirements to identify clearly those functions the proposed steam iron needed to perform, Company S systematized those functions.

Step 3: Systematize the functions. With the functions the steam iron needed to perform clearly identified, Company S organized the goal/means relationships among the functions and constructed a function family tree. Figure 7-22 shows the tree for the high-end steam iron.

On the function family tree, the ultimate goal or purpose of the iron was expressed by the function "straighten out clothes," and the function designated "smooth out wrinkles" was essential to that ultimate function. Although not expressed as a formal function expression, the function "crease clothing" was included here as well.

The "smooth out wrinkles" function, in turn, required four functions: "maintain base plate temperature," "generate steam," "exert pressure," and "form base plate."

The basic "smooth out wrinkles" function was not the only second-order function. There were a total of five functions

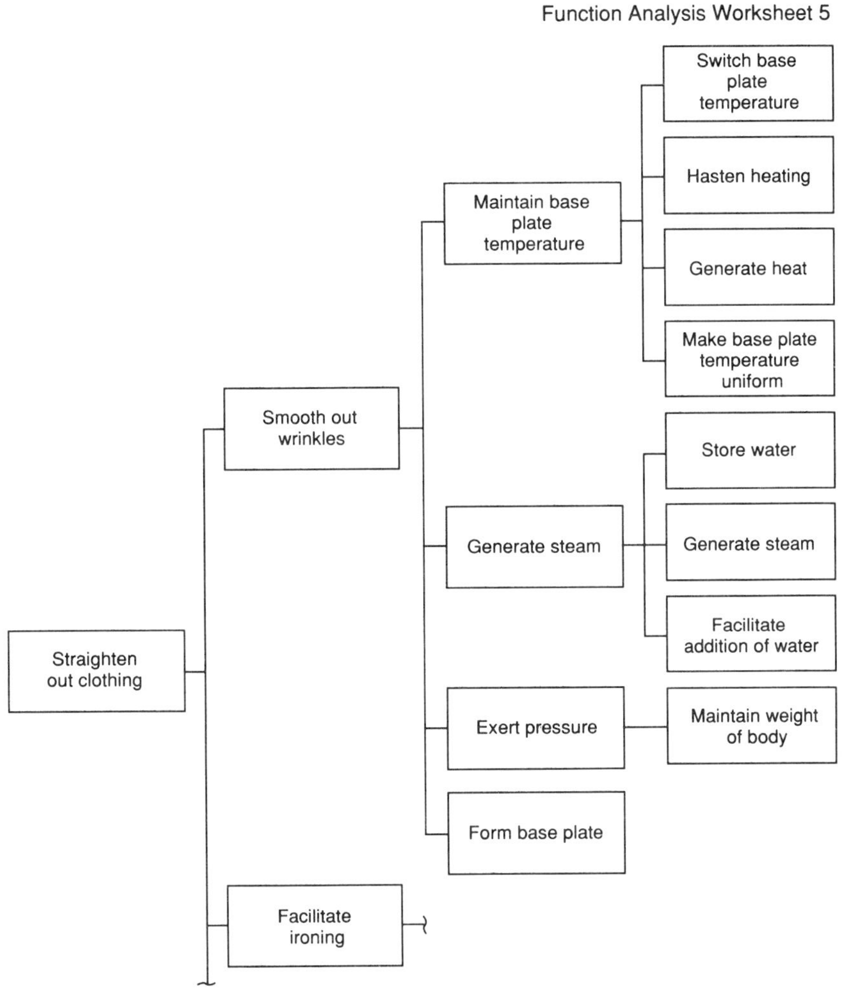

Figure 7-22. Function Family Tree for a Steam Iron

needed here. Besides the one just mentioned, there was "facilitate ironing" and three other functions. Omitted from the tree in our example were "facilitate storage," "assure iron safety," and "enhance iron attractiveness."

Completion of the function family tree clarified what the iron to be designed should be like (that is, showed its "design concept"). Next came specific design activities.

The first step in constructing a design draft was to determine the level of function targeted by design. In the case of product design, you make a design draft by examining all functions in all function fields. The company decided to focus on the four functions subordinate to the function field designated "smooth out wrinkles." Because designers could work out specific features of the iron by considering ways to achieve functions in the three function fields ("facilitate ironing," "facilitate storage," and "assure iron safety"), the company decided to make a thorough examination of those second-order functions. In other words, since basic iron principles were already established, Company S opted to develop a revolutionary new iron by investigating methods of achieving functions corresponding to functions that had been problems for irons in the past: storage, operability, safety, and so on.

In step 2, the company generated ideas for achieving each of the functions examined, evaluated those ideas, and summarized them concretely in the form of drawings.

Prototypes were built from the drawings in steps 3 and 4. These prototypes were checked in terms of function, quality, and cost, and then reflected in formal design drawings. We will leave out the details of the design drafts and only note that this process enabled the company to develop products characterized by enhanced operability and ease of storage. These new products included (1) a cordless steam iron and (2) an iron that could both steam and press (the ironing board having been given the function of generating steam).

Function Analysis Applied to the Design of a System for Receiving Visitors (Software)

Company K, a major construction firm, was rebuilding one of its branch offices to accommodate expanded operations.

Because visitor reception activities were important to business — serving as the face the company presented to the outside world — the firm decided to use function analysis to design an ideal system for receiving visitors.

Step 1: Understand the object of analysis. First, Company K identified the requirements of a visitor reception system and drew up a design theme description on the basis of those requirements. Figures 7-23(A) and 7-23(B) show this description.

Function Analysis Worksheet 2

Description of Support Activity Design Theme

1. Support system name
   Visitor reception system
2. Support system requirements
   Support activity goals and objectives
   Bring visitors and company hosts together smoothly (within 5 minutes)
   Present appealing image to visitors
   Identify undesirable visitors and contact General Affairs
   Departmental requirements for support system

	Executive Staff	Sales	Con-struction	General Affairs
Having General Affairs deal with undesirable visitors	○	○	○	○
Not keeping visitors waiting	○	○	○	○
Treating frequent visitors politely	×	×	○	○
Cutting the cost of reception activities	○	×	×	○
Improving the company's image	○	○	×	×
Using free time productively	○	×	×	○

○.......... (Important) ×.......... (Unimportant)

Figure 7-23(A). Description of Support Activity Design Theme

Function Analysis Worksheet 2 (continued)

3. Support system design requirements
   1) Rules and regulations
      - Reception task times of 8:45 AM to 5 PM conform to Labor Standards Act and company regulations
      - Reception activities take place on 1st floor
   2) Investment allowed
      $8,000 or less (including desks, chairs, telephones)
   3) Other
      - Use receptionists who will enhance the company image
      - Visitors include suppliers, consultants, representatives of other firms in the industry, customers, etc.

4. Support system running requirements
   1) Volume of work
      - Number of visitors per day: about 60 groups, 100 individuals
      - Peak hours of visits
        10:00 AM - 11:00 AM 30%
        1:00 PM - 2:00 PM 30%
        4:00 PM - 5:00 PM 20%
      - Other times 20%
   2) System and work quality
      - Time from visitor's arrival until meeting with company host should be 5 minutes or less
      - Treat all visitors with cheerful smiles
   3) Support activity personnel
      - Two people or fewer
   4) Other
      - Includes job of escorting visitors to reception room
      - Know beforehand of visitors with appointments
      - Contact appropriate company host by telephone for unexpected visitors

Figure 7-23(B). Support System Design Requirements

An examination of the requirements shown on the theme description reveals that tasks involved in receiving visitors are by nature difficult to quantify. For example, crucial requirements such as projecting a favorable image and greeting visitors with cheerful smiles resist quantification because they depend on "soft" aspects of human behavior. This feature is peculiar to support systems. It is still necessary, however, to pin down as many other requirements as possible by quantifying them.

In our example, the theme description clearly showed what sort of system should be designed, so Company K moved on to function definition.

Step 2: Define the functions. Function definition involves generating ideas for activities that will accomplish the goals and objectives of tasks identified in the theme description. In this case, Company K responded to the three goals and objectives by generating ideas for support activities. These activities were then evaluated and defined as functions. These processes of idea generation and function definition are shown in Figures 7-24 and 7-25.

In the case of the system for receiving visitors, task ideas generated for function definition had three goals. Ideas were proposed for all those goals. In our example, ideas addressing the three goals are listed randomly, but each idea was aimed at a particular goal. Although some of the ideas may seem obvious, it was important not to leave any gaps.

Company K next moved to evaluate each activity idea. The idea that the five employees abandon regular company uniforms in favor of their own clothing was rejected because of the increased burden of the cost of clothes and the likelihood of unevenness in dress among the employees. Also abandoned was item 9 — the suggestion that video cameras be installed to allow General Affairs to monitor visitors. This idea was rejected on the grounds of excessive initial investment and insufficient effectiveness.

After using various constraints and conditions on requirements to reduce the total pool of ideas, the company adopted more than 90 percent of the task ideas.

Next, Company K converted the ideas into expressions of functions that visitor reception activities needed to perform. For example, the idea of creating a favorable impression (activity 1) was expressed as the function of "give favorable impression." The idea of activity 2 — to contact the company host — was also expressed in a straightforward manner: "announce visitors."

Function Analysis Worksheet 3

Support System Name:	Visitor reception system
Work Goals and Objectives:	Bring visitors and company hosts together smoothly Present appealing image to visitors Identify undesirable visitors and contact General Affairs

Support Activities to Achieve Goals and Objectives	Evaluation
1. Make a favorable impression on visitors.	○
2. Contact company host immediately.	○
3. Escort visitors to reception room.	○
4. Identify visitors at reception desk and contact appropriate employee.	○
5. Have receptionists wear their own clothes, not company uniforms.	×
6. Have receptionists wear cheerful uniforms.	○
7. Occasionally note trends in visits and adjust tasks accordingly.	○
8. Redecorate the interior reception area to improve the atmosphere.	○
9. Install video cameras to allow General Affairs to monitor visitors.	×
10. Install chime button on reception desk so receptionist can contact General Affairs.	○
11. Provide employee attendance board to speed handling of visitors.	○
12. Provide company brochures at reception desk.	○
13. Review schedule of visits one day in advance.	○
14. Have visitors wear badges to distinguish them from company employees.	○
15. Establish and use criteria for identifying types of visitors.	○
16. Establish and use a blacklist.	○
17. Use outside personnel company to supply receptionists.	○
18. Make sure receptionists have adequate understanding of the company's business.	○
19. Issue three-year passes to frequent visitors and make sure that they use them.	○
20. Contact company hosts immediately when visitors arrive and have them met.	○
21. Have visitors show business cards when they arrive; check cards.	○
22. Determine the minimum expressions of welcome with which to greet visitors.	○

(etc.)

Figure 7-24. Generating Ideas for Support Activities

In this way, Company K defined each of the task ideas in terms of functions. It was crucial to undertake this process only after verifying the task idea's true significance.

In our example we have left out individual function criteria.

Function Analysis Worksheet 4

Support Activity	Functions		
	Verb	Noun	Criteria
1. Make a favorable impression on visitors	Give	Favorable impression	(omitted)
2. Contact company host immediately.	Announce	Visitors	"
3. Escort visitors to reception room.	Escort	Visitors	"
4. Identify visitors at reception desk and contact appropriate employee	Inform of Verify Judge	Visitor's arrival Visitor status Visitor acceptability	"
5. Have receptionists wear cheerful uniforms	Give Improve	Cheerful image Impression of reception	"
6. Occasionally note trends in visits and adjust tasks accordingly	Understand	Status of visits	"
7. Redecorate the interior reception area to improve the atmosphere	Improve	Reception image	"
8. Install chime button on reception desk to contact General Affairs	Announce	Visitors	"
9. Provide employee attendance board to speed handling of visitors	Know	Who is in	"
10. Make company brochures available at reception desk	Inform of	Business	"

(etc.)

Figure 7-25. Evaluating and Defining Support Activities as Functions

Step 3: Systematize the functions. Once the individual functions that a visitor reception system needed to perform had been clearly identified, Company K began organizing them according to their goal/means relationships. Figure 7-26 shows the function family tree for the visitor reception system.

Because the system functions as a contact point between the company and the outside world, it had to perform the function designated "carry out business smoothly." Satisfying this function basically required performing the function of "bring together visitors and hosts." Since visitors' impressions of the company are usually formed the minute they set foot inside the building, another extremely important function was "improve impression of company." The third-and fourth-order functions shown in the sample tree were needed for second-order functions like these.

The above procedure clearly and systematically identified what functions the reception system had to perform. Company K next moved on to the establishment of specific methods to perform those functions.

In the design-draft construction stage, Company K began by choosing a specific function level for generating ideas and examining concrete methods for the performance of individual visitor reception functions. Because second-order functions were too abstract for focusing in on concrete methods, in each function field the company examined ideas at the third level of functions.

Next, Company K generated ideas corresponding to each function and pared them down to concrete methods for performing reception system functions. The final design draft was organized as a job manual that was then used to train employees selected as receptionists.

This procedure produced a rational visitor reception system and an operations manual for carrying out the receptionist's job. The inclusion of the function family tree in the manual gave receptionists an adequate appreciation of the goals of their work. An important point for human-dependent work is the value of plotting jobs on a function family tree and using the tree to construct a job manual.

Function Analysis Worksheet 5

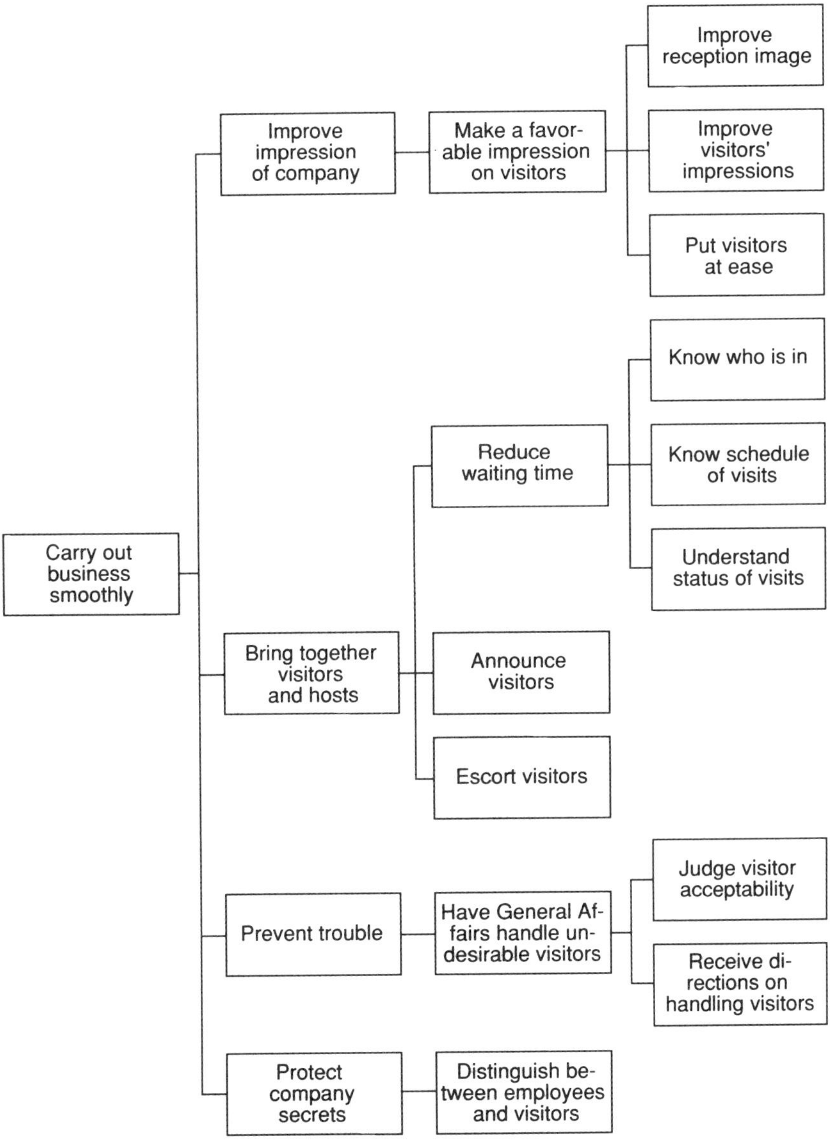

Figure 7-26. Function Family Tree for Visitor Reception System

References

Chapter 1

1. Masashige Sawada, *The Structure of Knowledge* (Nihon Hōsō Shuppan Kyōkai, 1969).
2. Sangyō Nōritsu University, VM Center, *VE Basics* (Sangyō Nōritsu University Press, 1986).

Chapter 2

1. Lawrence D. Miles, *Techniques of Value Analysis and Engineering* (New York: McGraw-Hill, 1961), p. 11-14.
2. Kaneo Akiyama, "FAST no kenkyū (1)," *Value Engineering* (Japan VE Association) 100 (1984):7.
3. Arthur E. Mudge, "How to Construct and Use a VE Function Chart," *Value Engineering* (1965).
4. Akiyama, "FAST no kenkyū(1)," p. 7.
5. Masatosi Tamai, "How to Build Functional Family Trees," *Value Engineering* 8 (1967):7.
6. Masayoshi Tanaka, *A Survey Concerning Target Costs, VE and Price Estimates in the Product Development and Design Process* (Japan VE Association, 1985).

Chapter 3

1. Kaneo Akiyama, "VE and Design," *Production and Transportation* Shingijutsusha 19 (1978), p. 5.
2. Kaneo Akiyama, "The TAP Method — An Integrated Approach to Process Design," *Annals of the Sangyō Nōritsu University Joint Institute* 16 (1985), p. 3.
3. Kaneo Akiyama, "A Study of Function Family Trees, Part 1," *Value Engineering* 94 (1983), p. 7.
4. Kaneo Akiyama, *Quality and Function,* vol. 2 of Sangyō Nōritsu University Joint Research Reports (1980).

Chapter 4

1. Material for the section on function definition was drawn from Sangyō Nōritsu University VM Center, ed., *VE Basics* (Sangyō Nōritsu University Press, 1986).
2. Kaneo Akiyama "A Study of Function Family Trees, Part 2," *Value Engineering* 95 (1987) p. 9.
3. Kaneo Akiyama and Hideharu Tanaka, "Value Analysis II," Sangyō Nōritsu University Correspondence Courses Division, 1979.
4. Kaneo Akiyama, "FAST Studies (3)," *Value Engineering* 102 (1984), p. 11.
5. Kaneo Akiyama, "FAST Studies (2)," *Value Engineering* 101 (1984) p. 9.
6. Kaneo Akiyama, "FAST Studies (8)," *Value Engineering* 114 (1986), p. 11.
7. Kaneo Akiyama, "FAST Studies (10)," *Value Engineering* 118 (1987), p. 7.
8. Kaneo Akiyama, "FAST Studies (6)," *Value Engineering* 110 (1986), p. 3.

Chapter 5

1. Shigeru Watanabe, *What Is a "System?"* (Kyōritsu Shuppan, 1974).

Chapter 6

1. Edward V. Krick, *An Introduction to Engineering and Engineering Design* (New York: John Wiley & Sons, 1969).

About the Author

Professor Kaneo Akiyama graduated in 1968 from Chiba Institute of Technology's Department of Industrial Management. He joined Sanno Junior College in 1975, where he served initially in its Business Management Center. Currently an assistant professor, he is section chief of the Value Development Center at Sanno College's Value Management Institute. His published books include *A Handbook for Inter-office Standardization* (Japan Standards Association), *Practical Procedures for Value Engineering Activities* (Keiei Jitsumu Press), and *The Basics of Value Engineering* (Sanno Press).

Index

OTHER BOOKS ON MANUFACTURING IMPROVEMENT

JIT Implementation Manual
The Complete Guide to Just-In-Time Manufacturing

by Hiroyuki Hirano

Here is the most comprehensive and detailed manual we have found anywhere for setting up a complete JIT program. Encyclopedic in scope, and written by a top international consultant, it provides the JIT professional with the answer to virtually any JIT problem. It shows multiple options for handling every stage of implementation and is appropriate to all factory settings, whether in job shop, repetitive, or process manufacturing. Covering JIT concepts, techniques, and tools, and including hundreds of illustrations, charts, diagrams, and JIT management forms, this manual is a truly indispensable tool.

ISBN 0-915299-66-6 / 1000+ pages in 2 volumes / $2500.00 / Order code HIRANO-BK

Kanban and Just-In-Time at Toyota
Management Begins at the Workplace (rev.)

Japan Management Association (ed.), David J. Lu (translator)

Based on seminars developed by Taiichi Ohno and others at Toyota for their major suppliers, this book is the best practical introduction to Just-In-Time available. Now in a newly expanded edition, it explains every aspect of a "pull" system in clear and simple terms — the underlying rationale, how to set up the system and get everyone involved, and how to refine it once it's in place. A ground breaking and essential tool for companies beginning JIT implementation.

ISBN 0-915299-48-8 / 224 pages / $36.50 / Order code KAN-BK

Canon Production System
Creative Involvement of the Total Workforce

compiled by the Japan Management Association

A fantastic success story! Canon set a goal to increase productivity by three percent per month — and achieved it! The first book-length case study to show how to combine the most effective Japanese management principles and quality improvement techniques into one overall strategy that improves every area of the company on a continual basis. Shows how the major QC tools are applied in a matrix management model.

ISBN 0-915299-06-2 / 251 pages / $36.95 / Order code CAN-BK

20 Keys to Workplace Improvement

by Iwao Kobayashi

This easy-to-read introduction to the "20 keys" system presents an integrated approach to assessing and improving your company's competitive level. The book focuses on systematic improvement through five levels of achievement in such primary areas as industrial housekeeping, small group activities, quick changeover techniques, equipment maintenance, and computerization. A scoring guide is included, along with information to help plan a strategy for your company's world class improvement effort.

ISBN 0-915299-61-5 / 264 pages / $34.95 / Order code 20KEYS-BK

Total Manufacturing Management
Production Organization for the 1990s

by Giorgio Merli

One of Italy's leading consultants discusses the implementation of Just-In-Time and related methods (including QFD and TPM) in Western corporations. The author does not approach JIT from a mechanistic orientation aimed simply at production efficiency. Rather, he discusses JIT from the perspective of industrial strategy and as an overall organizational model. Here's a sophisticated program for organizational reform that shows how JIT can be applied even in types of production that have often been neglected in the West, including custom work.

ISBN 0-915299-58-5 / 224 pages / $39.95 / Order code TMM-BK

Continuous Improvement in Operations
A Systematic Approach to Waste Reduction

Alan Robinson (ed.)

Now one handy book brings you the world's most advanced thinking on Just-In-Time, *kaizen*, Total Employee Involvement, and Total Productive Maintenance. Here in one volume is a compendium of materials from our best-selling classics by world-famous manufacturing experts. A lengthy introduction integrates the developments of these manufacturing gurus within a twofold theme the elimination of invisible waste and the creation of a work environment that welcomes and institutes employee's ideas. It's a perfect book for your study groups and improvement teams.

ISBN 0-915299-51-8 / 416 pages / $34.95 / Order ROB2C-BK

A Revolution in Manufacturing
The SMED System

by Shigeo Shingo, translated by Andrew P. Dillon

SMED (Single-Minute Exchange of Die), or quick changeover techniques, is the single most powerful tool for Just-In-Time production. Written by the industrial engineer who developed SMED for Toyota, the book contains hundreds of illustrations and photographs, as well as twelve chapter-length case studies. Here are the most complete and detailed instructions available anywhere for transforming a manufacturing environment to speed up production (Shingo's average setup time reduction is an astounding 98 percent) and make small-lot inventories feasible.
ISBN 0-915299-03-8 / 383 pages / $70.00 / Order code SMED-BK

A Study of the Toyota Production System
From an Industrial Engineering Viewpoint (rev.)

by Shigeo Shingo

The "green book" that started it all — the first book in English on JIT, now completely revised and re-translated. Here is Dr. Shingo's classic industrial engineering rationale for the priority of process-based over operational improvements for manufacturing. He explains the basic mechanisms of the Toyota production system in a practical and simple way so that you can apply them in your own plant.
ISBN 0-915299-17-8 / 294 pages / $39.95 / Order code STREV-BK

Variety Reduction Program (VRP)
A Production Strategy for Product Diversification

by Toshio Suzue and Akira Kohdate

Here's the first book in English on a powerful way to increase manufacturing flexibility without increasing costs. How? By reducing the number of parts within each product type and by simplifying and standardizing parts between models. VRP is an integral feature of advanced manufacturing systems. This book is both an introduction to and a handbook for VRP implementation, featuring over 100 illustrations, for top manufacturing executives, middle managers, and R&D personnel.
ISBN 0-915299-32-1 / 164 pages / $59.95 / Order code VRP-BK

COMPLETE LIST OF TITLES FROM PRODUCTIVITY PRESS

Akao, Yoji (ed.). **Quality Function Deployment: Integrating Customer Requirements into Product Design**
ISBN 0-915299-41-0 / 1990 / 387 pages / $ 75.00 / order code QFD

Asaka, Tetsuichi and Kazuo Ozeki (eds.). **Handbook of Quality Tools: The Japanese Approach**
ISBN 0-915299-45-3 / 1990 / 336 pages / $59.95 / order code HQT

Belohlav, James A. **Championship Management: An Action Model for High Performance**
ISBN 0-915299-76-3 / 1990 / 265 pages / $29.95 / order code CHAMPS

Birkholz, Charles and Jim Villella. **The Battle to Stay Competitive: Changing the Traditional Workplace**
ISBN 0-915-299-96-8 / 1991 / 110 pages / $9.95 /order code BATTLE

Christopher, William F. **Productivity Measurement Handbook**
ISBN 0-915299-05-4 / 1985 / 680 pages / $137.95 / order code PMH

D'Egidio, Franco. **The Service Era: Leadership in a Global Environment**
ISBN 0-915299-68-2 / 1990 / 165 pages / $29.95 / order code SERA

Ford, Henry. **Today and Tomorrow**
ISBN 0-915299-36-4 / 1988 / 286 pages / $24.95 / order code FORD

Fukuda, Ryuji. **CEDAC: A Tool for Continuous Systematic Improvement**
ISBN 0-915299-26-7 / 1990 / 144 pages / $49.95 / order code CEDAC

Fukuda, Ryuji. **Managerial Engineering: Techniques for Improving Quality and Productivity in the Workplace** (rev.)
ISBN 0-915299-09-7 / 1986 / 208 pages / $39.95 / order code ME

Grief, Michel. **The Visual Factory: Building Participation Through Shared Information**
ISBN 0-915299-67-4 / 1991 / 320 pages / $49.95 / order code VFAC

Hatakeyama, Yoshio. **Manager Revolution! A Guide to Survival in Today's Changing Workplace**
ISBN 0-915299-10-0 / 1986 / 208 pages / $24.95 / order code MREV

Hirano, Hiroyuki. **JIT Factory Revolution: A Pictorial Guide to Factory Design of the Future**
ISBN 0-915299-44-5 / 1989 / 227 pages / $49.95 / order code JITFAC

Hirano, Hiroyuki. **JIT Implementation Manual: The Complete Guide to Just-In-Time Manufacturing**
ISBN 0-915299-66-6 / 1990 / 1006 pages / $2500.00 / order code HIRANO

Horovitz, Jacques. **Winning Ways: Achieving Zero-Defect Service**
ISBN 0-915299-78-X / 1990 / 165 pages / $24.95 / order code WWAYS

Japan Human Relations Association (ed.). **The Idea Book: Improvement Through TEI (Total Employee Involvement)**
ISBN 0-915299-22-4 / 1988 / 232 pages / $49.95 / order code IDEA

Japan Human Relations Association (ed.). **The Service Industry Idea Book: Employee Involvement in Retail and Office Improvement**
ISBN 0-915299-65-8 / 1990 / 294 pages / $49.95 / order code SIDEA

Japan Management Association (ed.). **Kanban and Just-In-Time at Toyota: Management Begins at the Workplace** (rev.), Translated by David J. Lu
ISBN 0-915299-48-8 / 1989 / 224 pages / $36.50 / order code KAN

Japan Management Association and Constance E. Dyer. **The Canon Production System: Creative Involvement of the Total Workforce**
ISBN 0-915299-06-2 / 1987 / 251 pages / $36.95 / order code CAN

Jones, Karen (ed.). **The Best of TEI: Current Perspectives on Total Employee Involvement**
ISBN 0-915299-63-1 / 1989 / 502 pages / $175.00 / order code TEI

JUSE. **TQC Solutions: The 14-Step Process**
ISBN 0-915299-79-8 / 1991 / 416 pages / 2 volumes / $120.00 / order code TQCS

Kanatsu, Takashi. **TQC for Accounting: A New Role in Companywide Improvement**
ISBN 0-915299-73-9 / 1991 / 244 pages / $45.00 / order code TQCA

Karatsu, Hajime. **Tough Words For American Industry**
ISBN 0-915299-25-9 / 1988 / 178 pages / $24.95 / order code TOUGH

Karatsu, Hajime. **TQC Wisdom of Japan: Managing for Total Quality Control**, Translated by David J. Lu
ISBN 0-915299-18-6 / 1988 / 136 pages / $34.95 / order code WISD

Kaydos, Will. **Measuring, Managing, and Maximizing Performance**
ISBN 0-915299- 98-4 / 1991 / 208 pages / $34.95 / order code MMMP

Kobayashi, Iwao. **20 Keys to Workplace Improvement**
ISBN 0-915299-61-5 / 1990 / 264 pages / $34.95 / order code 20KEYS

Lu, David J. **Inside Corporate Japan: The Art of Fumble-Free Management**
ISBN 0-915299-16-X / 1987 / 278 pages / $24.95 / order code ICJ

Merli, Giorgio. **Total Manufacturing Management: Production Organization for the 1990s**
ISBN 0-915299-58-5 / 1990 / 224 pages / $39.95 / order code TMM

Mizuno, Shigeru (ed.). **Management for Quality Improvement: The 7 New QC Tools**
ISBN 0-915299-29-1 / 1988 / 324 pages / $59.95 / order code 7QC

Monden, Yasuhiro and Michiharu Sakurai (eds.). **Japanese Management Accounting: A World Class Approach to Profit Management**
ISBN 0-915299-50-X / 1990 / 568 pages / $59.95 / order code JMACT

Nachi-Fujikoshi (ed.). **Training for TPM: A Manufacturing Success Story**
ISBN 0-915299-34-8 / 1990 / 272 pages / $59.95 / order code CTPM

Nakajima, Seiichi. **Introduction to TPM: Total Productive Maintenance**
ISBN 0-915299-23-2 / 1988 / 149 pages / $39.95 / order code ITPM

Nakajima, Seiichi. **TPM Development Program: Implementing Total Productive Maintenance**
ISBN 0-915299-37-2 / 1989 / 428 pages / $85.00 / order code DTPM

Nikkan Kogyo Shimbun, Ltd./Factory Magazine (ed.). **Poka-yoke: Improving Product Quality by Preventing Defects**
ISBN 0-915299-31-3 / 1989 / 288 pages / $59.95 / order code IPOKA

Ohno, Taiichi. **Toyota Production System: Beyond Large-Scale Production**
ISBN 0-915299-14-3 / 1988 / 162 pages / $39.95 / order code OTPS

Ohno, Taiichi. **Workplace Management**
ISBN 0-915299-19-4 / 1988 / 165 pages / $34.95 / order code WPM

Ohno, Taiichi and Setsuo Mito. **Just-In-Time for Today and Tomorrow**
ISBN 0-915299-20-8 / 1988 / 208 pages / $34.95 / order code OMJIT

Perigord, Michel. **Achieving Total Quality Management: A Program for Action**
ISBN 0-915299-60-7 / 1991 / 384 pages / $45.00 / order code ACHTQM

Psarouthakis, John. **Better Makes Us Best**
ISBN 0-915299-56-9 / 1989 / 112 pages / $16.95 / order code BMUB

Robinson, Alan. **Continuous Improvement in Operations: A Systematic Approach to Waste Reduction**
ISBN 0-915299-51-8 / 1991 / 416 pages / $34.95 / order code ROB2-C

Robson, Ross (ed.). **The Quality and Productivity Equation: American Corporate Strategies for the 1990s**
ISBN 0-915299-71-2 / 1990 / 558 pages / $29.95 / order code QPE

Shetty, Y.K and Vernon M. Buehler (eds.). **Competing Through Productivity and Quality**
ISBN 0-915299-43-7 / 1989 / 576 pages / $39.95 / order code COMP

Shingo, Shigeo. **Non-Stock Production: The Shingo System for Continuous Improvement**
ISBN 0-915299-30-5 / 1988 / 480 pages / $75.00 / order code NON

Shingo, Shigeo. **A Revolution In Manufacturing: The SMED System**, Translated by Andrew P. Dillon
ISBN 0-915299-03-8 / 1985 / 383 pages / $70.00 / order code SMED

Shingo, Shigeo. **The Sayings of Shigeo Shingo: Key Strategies for Plant Improvement**, Translated by Andrew P. Dillon
ISBN 0-915299-15-1 / 1987 / 208 pages / $39.95 / order code SAY

Shingo, Shigeo. **A Study of the Toyota Production System from an Industrial Engineering Viewpoint** (rev.)
ISBN 0-915299-17-8 / 1989 / 293 pages / $39.95 / order code STREV

Shingo, Shigeo. **Zero Quality Control: Source Inspection and the Poka-yoke System**,Translated by Andrew P. Dillon
ISBN 0-915299-07-0 / 1986 / 328 pages / $70.00 / order code ZQC

Shinohara, Isao (ed.). **New Production System: JIT Crossing Industry Boundaries**
ISBN 0-915299-21-6 / 1988 / 224 pages / $34.95 / order code NPS

Sugiyama, Tomo. **The Improvement Book: Creating the Problem-Free Workplace**
ISBN 0-915299-47-X / 1989 / 236 pages / $49.95 / order code IB

Suzue, Toshio and Akira Kohdate. **Variety Reduction Program (VRP): A Production Strategy for Product Diversification**
ISBN 0-915299-32-1 / 1990 / 164 pages / $59.95 / order code VRP

Tateisi, Kazuma. **The Eternal Venture Spirit: An Executive's Practical Philosophy**
ISBN 0-915299-55-0 / 1989 / 208 pages/ $19.95 / order code EVS

Yasuda, Yuzo. **40 Years, 20 Million Ideas: The Toyota Suggestion System**
ISBN 0-915299-74-7 / 1991 / 210 pages / $39.95 / order code 4020

Audio-Visual Programs

Japan Management Association. **Total Productive Maintenance: Maximizing Productivity and Quality**
ISBN 0-915299-46-1 / 167 slides / 1989 / $749.00 / order code STPM
ISBN 0-915299-49-6 / 2 videos / 1989 / $749.00 / order code VTPM

Shingo, Shigeo. **The SMED System**, Translated by Andrew P. Dillon
ISBN 0-915299-11-9 / 181 slides / 1986 / $749.00 / order code S5
ISBN 0-915299-27-5 / 2 videos / 1987 / $749.00 / order code V5

Shingo, Shigeo. **The Poka-yoke System**, Translated by Andrew P. Dillon
ISBN 0-915299-13-5 / 235 slides / 1987 / $749.00 / order code S6
ISBN 0-915299-28-3 / 2 videos / 1987 / $749.00 / order code V6

Returns of AV programs willl be accepted for incorrect or damaged shipments only.

TO ORDER: Write, phone, or fax Productivity Press, Dept. BK, P.O. Box 3007, Cambridge, MA 02140, phone 1-800-274-9911, fax 617-864-6286. Send check or charge to your credit card (American Express, Visa, MasterCard accepted).

U.S. ORDERS: Add $5 shipping for first book, $2 each additional for UPS surface delivery. CT residents add 8% and MA residents 5% sales tax. For each AV program that you order, add $5 for programs with 1 or 2 tapes, and $12 for programs with 3 or more tapes.

INTERNATIONAL ORDERS: Write, phone, or fax for quote and indicate shipping method desired. Pre-payment in U.S. dollars must accompany your order (checks must be drawn on U.S. banks). When quote is returned with payment, your order will be shipped promptly by the method requested.

NOTE: Prices subject to change without notice.